Behind the Cloud

Voices About the Book

"Goosebumps guaranteed: Reads like a thriller but is, unfortunately, a very realistic insight into the "brave new world" of digitalization. Since Edward Snowden, we know that there is basically no privacy left that would not be spied on—and not only in China. From Peter Seele and Lucas Zapf, we learn knowledgeably how sophisticated this is now happening".
—Stephan Russ-Mohl, *media researcher, founder of the European Journalism Observatory*

"Privacy, its subject matter and its protection, must be redefined in the context of digitalisation, not least in legal terms. Peter Seele and Lucas Zapf provide a knowledgeable and convincing account of what is at stake here".
—Julian Krüper, *Chair of Public Law, Constitutional Theory and Interdisciplinary Legal Research, Ruhr University Bochum (Verfassungstheorie und interdisziplinäre Rechtsforschung, Ruhr Universität Bochum)*

Peter Seele • Lucas Zapf

Behind the Cloud

A Theory of the Private Without Secrecy

Peter Seele
Università della Svizzera italiana
Lugano, Switzerland

Lucas Zapf
University of Basel
Basel, Switzerland

This book is a translation of the original German edition „Die Rückseite der Cloud" by Seele, Peter, published by Springer-Verlag GmbH Germany in 2020. The translation was done with the help of artificial intelligence (machine translation by the service DeepL.com). A subsequent human revision was done primarily in terms of content, so that the book will read stylistically differently from a conventional translation. Springer Nature works continuously to further the development of tools for the production of books and on the related technologies to support the authors.

ISBN 978-3-662-64501-7   ISBN 978-3-662-64502-4 (eBook)
https://doi.org/10.1007/978-3-662-64502-4

© Springer-Verlag GmbH Germany, part of Springer Nature 2022

This work is subject to copyright. All rights are reserved by the Publisher, whether the whole or part of the material is concerned, specifically the rights of reprinting, reuse of illustrations, recitation, broadcasting, reproduction on microfilms or in any other physical way, and transmission or information storage and retrieval, electronic adaptation, computer software, or by similar or dissimilar methodology now known or hereafter developed.

The use of general descriptive names, registered names, trademarks, service marks, etc. in this publication does not imply, even in the absence of a specific statement, that such names are exempt from the relevant protective laws and regulations and therefore free for general use.

The publisher, the authors and the editors are safe to assume that the advice and information in this book are believed to be true and accurate at the date of publication. Neither the publisher nor the authors or the editors give a warranty, expressed or implied, with respect to the material contained herein or for any errors or omissions that may have been made. The publisher remains neutral with regard to jurisdictional claims in published maps and institutional affiliations.

This Springer imprint is published by the registered company Springer-Verlag GmbH, DE part of Springer Nature.
The registered company address is: Heidelberger Platz 3, 14197 Berlin, Germany

"Thoughts are free" (Die Gedanken sind frei) goes the old folk saying—but thoughts are no longer secret, according to this book's thesis. Whereas in the past, it was (a) omniscient deities who knew everything about us, now, after a brief period of (b) secret privacy, our most secret characteristics, desires, and inclinations can once again be read out thanks to Big Data: the private secrets are stored on the "back of the cloud" and are evaluated there, and at the same time they are removed from our access. The knowledge of their visibility, the lifting of their secrecy, and their systematic use have an impact on our understanding of the private—and the independence of our thoughts themselves. We live in the age of (c) privacy without secrecy. In this book, the three types of the private are developed over the course of history into a new theory of the private without secrets, using numerous example cases, such as Airbnb, Uber, Parship, or Hello Barbie.

Contents

1 **Introduction: Behind the Cloud** — 1
  1.1 On the Changing Relationship Between Privacy and Secrecy in the Digital Age — 1
  1.2 Facets of Change: Privacy Without Secrecy on Three Levels — 3
  1.3 Digital Promises: Masters of the Universe with Secret Weaknesses — 6
  1.4 Distinction from Cases Not Dealt with — 9
  1.5 Does the Digital Age Still Need Theory? — 11
  References — 12

Part I The Secret Private: Introduction and Derivation — 15

2 **"Privacy Is Dead": How Could It Come to This?** — 17
  2.1 What Is the Private Sphere? — 18
  2.2 The Secret Private: Three Types — 21
    2.2.1 God Sees Everything: Transcendent Analogous Omniscience — 23
    2.2.2 "Secret Privates": Immanent Analogical Self-Knowledge — 25
    2.2.3 "Privacy Without Secrecy": Immanent Digital Omniscience — 28
    2.2.4 Encroachments on the Secret Private in Digital Omniscience — 34
  2.3 Mass Society as a Social Precondition of Immanent-Digital Omniscience — 35

	2.3.1	Historical Materialism and De-individualization	36
	2.3.2	The Drive towards Individualisation: Counterculture, *Rebel Sell* and Digitalisation	39
	2.3.3	Summary: Digital, Individualised Mass Society and the Abolition of the Secret Private Sphere	42
References			46

Part II Symptoms of the Structural Change of the Private   51

3   Showcasing Digital Omniscience in Everyday Life   53
   3.1 Driving a Taxi   53
       3.1.1 "Hello Taxi!"   53
       3.1.2 TaxiApp   54
       3.1.3 Uber   56
   3.2 Overnight Stay   57
       3.2.1 The Middle-Class Bedroom   58
       3.2.2 Overnight Stay in the Boarding House   58
       3.2.3 Airbnb   59
   3.3 Celebrating and Eating   60
       3.3.1 'Everyone Brings Something'   61
       3.3.2 Running Dinner   61
       3.3.3 Food and Party   62
   3.4 Sharing   66
       3.4.1 St Martin: Sharing out of Religious Conviction   66
       3.4.2 Collaborative Consumption: Sharing for a Better World   68
       3.4.3 Sharing Economy: Sharing as a Business Case   70
   3.5 Tying up   72
       3.5.1 The Village Fete   72
       3.5.2 Speed Dating   73
       3.5.3 Parship   74
   3.6 Advertising and Recommendations   76
       3.6.1 Billboard, Newspaper Advertisement and Personal Recommendation   77
       3.6.2 Quota Boxes and Direct Marketing   79
       3.6.3 Integrated Personalised Advertising: AdWorks and Spying Billboards   81
   3.7 Surveillance   84
       3.7.1 The Village Policeman   85

	3.7.2	Video Surveillance/CCTV	86
	3.7.3	Widespread Access to Private Communications: General Surveillance	88
3.8	Work and Employment		90
	3.8.1	Natural Working Rhythm	91
	3.8.2	The Time Clock	92
	3.8.3	Smartphone Tracking by the Boss	93
3.9	Election and Political Advertising		95
	3.9.1	The Secret Ballot and the Election Poster	96
	3.9.2	Voting Machines and Civil Dialogue	97
	3.9.3	Obama and Pandora	98
3.10	Networks		100
	3.10.1	Pinboard	100
	3.10.2	Analogue-Digital Information Networks	102
	3.10.3	The Powerful Digital Network	103
3.11	Payment and Digital Currencies		106
	3.11.1	Cash	106
	3.11.2	Credit Card	107
	3.11.3	Cryptocurrency	109
3.12	Books and e-Books		110
	3.12.1	One Edition, One Word	111
	3.12.2	Zeros and Ones Are Patient: Books and e-Books in Peaceful Co-Existence	112
	3.12.3	E-Books: When the Reader Reads the Reader	113
3.13	Sexuality and the Internet: The Incognito Illusion		114
	3.13.1	Adult Entertainment from the Station Bookshop	115
	3.13.2	When Pictures Learned to Surf	116
	3.13.3	The Incognito Illusion, Fitness Trackers and Bedside Bugs	116
References			123

Part III   Theory of a Structural Change of the Private   131

4   The Private Sphere Changes: A Consequence of Digitalization   133
   4.1   Economic Structural Change of the Private Sector   134
      4.1.1   Political and Personal Significance of the Economy and Its Digitalisation   135
      4.1.2   Economic Use of Personal Information   138

	4.1.3	The Secret Private as a Business Case: Seamless Products and Platform Capitalism	140
	4.1.4	The Shaping of the Secret Private by Companies	143
4.2	Political Structural Change of the Private Sector		144
	4.2.1	Intrusion of Politics into the Secret Private Sphere of Citizens	145
	4.2.2	Mixing Politics and Economics Through the Use of the Secret Private Sphere	148
	4.2.3	Opposition to the Political Domination of the Private Sphere	150
	4.2.4	Democratic-Legislative Updating of the Concept of Privacy	153
4.3	Social Structural Change of the Private Sphere		155
	4.3.1	More Exchange, Less Self-Determination: Informational Heteronomy	156
	4.3.2	New Social Spaces: Digital Intentionality and Self-Policing	159
		4.3.2.1 Digital Intentionality	160
		4.3.2.2 Self-Policing Instead of the Right to Be Forgotten	162
References			164

5 Summary: Thoughts in a Digital World: Free, but no Longer Secret — 169
- 5.1 Typology of the Secret Private — 170
- 5.2 Symptoms and Theory of Structural Change — 171
- 5.3 Digital Formation of the Secret Private — 173
- References — 174

6 Conclusion: Our Secrets Behind the Cloud — 175
- 6.1 The Updated Concept of the Secret — 175
- 6.2 The Secret Private in the Realm of Machines — 176
- 6.3 What to Do? — 178
  - 6.3.1 Making People Aware of What They Have Made — 179
  - 6.3.2 Conscious Use of the Digital Infrastructure — 179
  - 6.3.3 Privacy as a Business Model — 180
- 6.4 In Conclusion — 182
- References — 182

7	**Outlook: Digital Authenticity: Immersive Consumption Without Secrets**	**185**
	7.1 What Is Authenticity?	186
	7.2 Disney and Audi: Authenticity Brings Sales	188
	7.3 Authenticity Without Secrecy	189
	References	189

From Ethical Considerations to Proposed Legal Solutions: An Afterword by Bertil Cottier   191

List of Figures

Fig. 2.1	Is there a secret private and who knows about it? Three types	22
Fig. 2.2	Analogous places of the secret private	29
Fig. 2.3	Digital places of the secret private	30
Fig. 2.4	Encroachment on the individual in digital information processing	35
Fig. 2.5	Social preconditions of mass society	45
Fig. 3.1	St. Martin of Tours at Basel Cathedral. (Photo: Meskens 2010)	67
Fig. 3.2	Times Square in New York City and its billboards. (Photo: Benoist 2012)	77
Fig. 3.3	Threat and video surveillance. Own photograph clz, 01.06.2016, Basel-Stadt	87
Fig. 3.4	The three phases of the Internet's erosion and distortion of privacy	118
Fig. 4.1	Points of crystallization of the theory of the private without secrets in economics, politics and social affairs	134
Fig. 4.2	Declaration of war on the surveillance state. Own photograph clz, 27.04.2016, Lugano/TI	153
Fig. 4.3	Rejection of surveillance in public spaces. Own photograph clz, 01.06.2016, Allschwil BL	163
Fig. 5.1	Summary of the three types of secret privacy	171
Fig. 5.2	Structural change of the private sphere on the economic, political and social levels	173

1

Introduction: Behind the Cloud

1.1 On the Changing Relationship Between Privacy and Secrecy in the Digital Age

Behind the cloud is by no means empty and uninhabited. Whoever is there has insight into the private data of those who have entrusted themselves to the cloud. Our data is entrusted to the Internet, collected both with and without our consent. This book is about the transition that our understanding of privacy is going through and has already gone through. That which we assumed was protected by the principle of privacy is no longer inaccessible and thus secret. In order to describe this change of privacy and to put it into a new theory, in this book we deal with the connection between privacy and secrecy—and how this connection has been shaken by digitalization.

The trail of data that you constantly leave, almost seamlessly records who was when, where and in whose presence. Possibly even in what state of mind. Anyone carrying a mobile phone is tracked by the triangulation of transmission masts, WLAN logging or the storage of GPS geodata. Private vehicles are registered by license plate scanners. Mobile phones, laptops, game consoles and even car rear-view mirrors are equipped with cameras that detect whether someone is sleeping (microsleep in the car, for example), smiling or yawning. The collected data is compiled into profiles that enable personalized content and advertising, and the effectiveness of said advertising can be measured and assessed.

The translation of this chapter was done with the help of artificial intelligence (machine translation by the service DeepL.com). A subsequent human revision was done primarily in terms of content.

© Springer-Verlag GmbH Germany, part of Springer Nature 2022
P. Seele, L. Zapf, *Behind the Cloud*, https://doi.org/10.1007/978-3-662-64502-4_1

There are few limits to individual tracking based on seemingly impersonal data parameters. The site *Panopticlick*, for example, shows every Internet user how easy it is to create a unique profile of the website visitor based only on the publicly accessible browser and Internet settings *(canvas fingerprinting)*. In most cases, the configuration is unique. Just by the way we have moved a mouse pointer over the screen in the past, an identity can be established. Indeed, a useful service when it comes to your own online banking transactions. It is remarkable, however, when companies or state actors who pursue their interests with the individually assignable data are behind the analysis. It becomes dramatic and dangerous if this data falls into the wrong hands—reprisals or discrimination could ensue.

In addition to the wealth of personal data generated, it is especially algorithms that can recognize patterns from large amounts of data (Big Data) and even make predictions. The case of a young woman who was offered pregnancy products before she herself knew she was pregnant is now several years old and represents only the tip of the iceberg. Besides the legal side of using private data, there is the illegal side. Time and time again, data leaks dominate the news. Anyone who has given sensitive data to a company such as a bank, a dating platform or a credit card company in order to request a service has occasionally had to wonder or have been worried about their private data being stolen and published. The small and large secrets of the individual are thus read out by third parties with and without the consent of the individual. So in the digital age, the question is not: *Who am I—and if so, how many?* (Precht 2007) It is: *Who knows me—and if so, how many?*

Our main argument is that this brave new world of data is leading to a structural change in the private sphere. As the secret disappears from the private, the structural change takes place. This is not just about personal conscious knowledge that is kept secret. It is also about *unconscious* information which, as meta-data, reveals much about the individual, thus belonging, as it were, to the ego and forming part of the individual secret. This condition, however, is a new one: no sooner did the individual secret appear to have been fought for and secured than it was lost again. For a private, secret in the sense that only the person in question has access to it and of which others know nothing, is a novelty in the history of ideas. According to our typology of concepts of privacy, which is oriented towards the epochs, it only began with secularization. In pre-modern times, people kept their small and large secrets from each other, but not from God, if we take the Abrahamic cultural sphere as an example. God's omniscience encompassed the entire private sphere and thus every conscious and unconscious secret. Only with secularization did man gain supremacy and control over his secret. And with digitalization, he is losing that control again. This and the effects on the social, the political and the economic are the subject of the following book, The *Structural Change of the Private (Strukturwandel des Privaten)*.

1.2 Facets of Change: Privacy Without Secrecy on Three Levels

In the individual lifeworld, digital media cause profound changes: they have an effect personally, on values, politically, economically, touching every aspect of human action (cf. Heffernan 2016). Thanks to their access to the private, these digital media not only have the possibility to use this private, but also to shape it. This shaping is reminiscent of Marshall McLuhan's statement from the 1960s. McLuhan worried about the transformation of communication through electronics, especially television:

> All media work us over completely. They are so pervasive in their personal, political, economic, aesthetic, psychological, moral, ethical, and social consequences that they leave no part of us untouched, unaffected, unaltered (McLuhan et al. 2005, p. 26).

The digital age and the associated transformation in media and communication are changing all areas of life. The secret private sphere is particularly affected, because it is stored in the digital realm and thus accessible in principle. This is true even for those who do not yet have a concept of private and public, such as small children. The *Hello Barbie toy doll,* launched in 2015, vividly illustrates this potential, technology-driven public sphere of formerly private secrets. *Hello Barbie* is equipped with a microphone, a speaker, and a Wi-Fi interface, making it unlike earlier toy dolls that had, at most, a handful of sentences stored on a chip. *Hello Barbie* can record speech, send it to a server, evaluate the conversations—and respond (cf. Neumann 2015). Barbie—or rather her manufacturer—thus gains access to the conversations that are held with and around the doll. But that's not all: as one learns, the conversation is not only recorded, but also transcribed. If desired, the transcript will be sent to the parents via email in the next step. *Big Barbie is watching you.*

With *Hello Barbie,* the legal guardians consent to the monitoring. However, when adult, consent must then be given by the persons themselves. This is apparently done explicitly and often tacitly, as can be seen from the enormous global spread and extensive use of digitalised services. People under 35 spend an international average of over 2 h a day on social networks. The Facebook offerings alone (Facebook, Instagram and WhatsApp) report a combined 2.8 billion registered accounts (cf. Kroker 2015a, b). In view of these figures, it seems difficult to escape the digital offers attacking privacy. They exert a siren-like attraction on people, as Lanier (2014) pointed out with the term 'siren servers'. Just as it once was for Ulysses, it is now up to the users to avoid succumbing to the beckoning sounds. The sirens' promises sound familiar. Already in the Twelfth Canto of the Odyssey it fluted:

> Come, Ulysses, you pride of the Achaians, steer the ship to the shore to listen to our song! No one has passed the island in the dark ship without enjoying the sweet sounds from our mouth, then continued the journey satisfied and richer in knowledge. Well-informed we are of all that Greeks and Trojans suffered at the will of the gods in wide Troy, know also otherwise what all happens on the nourishing earth (Homer and Ebener 1976).

Sirens lure sailors and surfers alike by playing on human curiosity. Knowing more than before, knowing more than others, straight from the source. Odysseus knows the tragic outcome for those who succumb to the sirens' temptation. That is why he allows himself to be bound. He exposes himself to the song, wanting to experience the luring calls, but orders his men to cover their ears and not to untie him under any circumstances. And yet, as soon as he heard the first luring sounds, he wants to be untied. How do we deal with the luring sounds of today's sirens? What danger do they pose, and who holds back those who succumb to the lures with their eyes open?

Due to the proliferation of digital-electronic and sensor-enabled devices, such as smartphones, the temptations of siren servers are omnipresent and the systematic spying on privacy thereby the rule (cf. e.g. Spehr 2015): in addition to *personal devices*, it is cameras in public places through whose data collection the private can no longer be maintained as personal and secret (cf. Tryfonas et al. 2016; Schneider 2015, p. 2 f.).

In public and semi-public spaces (shops, apartment complexes or businesses), cameras record our actions and record the collected information in governmental and non-governmental databases. Thus, the sanctuaries of the secret private become fewer and fewer, the amount of information constantly increases and is used powerfully and effectively by pattern-recognizing algorithms. An awareness is emerging that everything we do can always become public: The digital does not forget, private things are publicly discoverable. A humorous side note sums up this awareness of the potential public: "Got my finger stuck in the garlic press and can't get it out. I'd love to google 'finger stuck in garlic press what to do' right now—but I'd rather not. I'm too embarrassed even in front of the algorithms" (Werner 2015). The algorithms are admittedly not the personal instance in front of which to feel embarrassment and shame. Rather, it is the patterns and data storage of all those requests that are keyworded and evaluated for consumption profiles that subsequently cause hesitancy and distance. What seems amusing in this example takes on an oppressive dimension when we apply it to the voting behaviour of politicians. For example, it is reported that the introduction of electronic voting among parliamentarians in the Swiss parliament led to politicians voting differently when they thought they were being digitally observed (cf. Bütler 2015). Not that the vote can be reconstructed, but behaviour is apparently fundamentally adjusted when exposed to digital surveillance.

The attacks on the secret do not stop at misadventures or parliamentary work. They also take place close to people, in their private lives, within their own four walls. The social network provider Facebook, for example, takes the liberty of accessing the microphone of the device during status entry and checking which music or which television programme is being listened to or watched while writing (cf. Facebook 2014). These developments are part of the larger context of lifeworld changes brought about by digitalization. As the digital becomes part of reality, the technical categories online, offline, real, virtual and their respective lifeworld implications become part of a new understanding of the world (cf. Floridi 2014; Ribi 2016). The private is severely affected by this change. Our hypothesis is therefore:

> Privacy must be redefined in the digital age.
> It exists only without secrecy.

Through this "theory of privacy without secrecy" we contribute to a realistic assessment of the current nature of privacy. The starting point of this hypothesis is a pre-digital notion of privacy, understood as the assured absence of judgments and measurements about the individual or the availability of informational bases for these purposes. In a highly digitalized and networked society, however, such an understanding of privacy seems obsolete: the technical structures are continuously present, measuring and surveying, judging and storing our actions, even making future actions predictable through pattern-recognizing algorithms. Nevertheless, this does not imply the abolition of the private sphere. Legally, the private sphere is still alive and relevant. The private still exists when the door to the bedroom is locked and the shutter is lowered.

But as long as the smartphone is on the bedside table, it is a private without a secret.

The following descriptions deal with the effects of the digital age on individual privacy on three levels: economic, social and political. The structural transformation of the private sphere is taking place on these three levels, as the following examples illustrate:

- *Economics*—Eric Schmidt, former CEO of Google: "We know where you are—with your permission. We know where you've been—with your permission. We can more or less know what you're thinking about" (Schmidt 2010). And the search engine knows—Schmidt does not mention this specifically—how to make money from the knowledge of its users' secrets. The effects of the private without the secret offer completely new business opportunities.
- *Social using "foodporn"* as *an example*—Sean Garrett, social media expert: "Early Twitter: 'What I had for lunch.' Early Meerkat: 'Watch me eat my lunch'" (Garrett 2015). Every mundane aspect of private life is becoming a digital event. With the "Meerkat" platform, it even becomes a video stream in real time—with or without viewers, but archived in any case. Anyone who keeps their private life to themselves quickly becomes an eccentric—*privacy is theft*, writes Dave Eggers (on this in detail Atwood 2013). According to the motto: Those who keep their private things to themselves rob the public of their shared experience.
- *Politics*—David Cameron, Prime Minister of Great Britain at the time: "In our country, do we want to allow a means of communication between people which [...] we cannot read? (Cameron 2015) In the impetus of the protective statesman, he asks this rhetorical question—the negation of which would abolish privacy protected from state access in the blink of an eye.

The structural change of the private sphere is built on these three levels, and it is on these levels that we base our theory to be developed here.

1.3 Digital Promises: Masters of the Universe with Secret Weaknesses

'Digitalization' is a buzzword that characterizes social change. We concretize digitalization with regard to the topics of secrecy, privacy and data protection and observe the changes in the secret private sphere on the three levels of economy, social and politics. Before we turn to these social anchors, we first ask how 'digitization' could achieve such a broad social impact.

In addition to the widespread dissemination and widespread use of digital offerings from Apple to Zalando, individual, particularly visible players in the industry are playing a decisive role in the dovetailing of digitization and society. Thanks to media attention and legions of interested users, CEOs like Mark Zuckerberg (Facebook) or Tim Cook (Apple) are becoming international, socially relevant players who keep themselves in the conversation with their ideas and innovations. In doing so, they are not just acting as entrepreneurs. They leave the sphere of the economy and become social influencers to whom political weight is attributed. The industry takes on a leading social function.

Who attracts the best graduates? Which company suggests to the employee that 'you've made it', conveys values such as success and progress? Those industries that are considered flagships and models of success, movers and shakers, are tied to social trends. The interplay of economic leadership is particularly visible in the US, which, as the world's largest economy, is at the forefront of this change. In the 1960s, it was industrial conglomerates such as General Electric and GM, producers of tangible consumer goods, that were regarded as the hub of the business world. A few years later, these were replaced by the finance and investment sector. The smart banker who generates millions in a fraction of a second with his computer. And today it is companies like Amazon, Alphabet and Facebook that stand for successful entrepreneurship worldwide with their range of services (cf. Henkel 2019 for US development and Dietz 2014 on the change in values). It is no coincidence, then, that precisely the three companies mentioned, Amazon, Alphabet and Google, were the top 3 most attractive employers in the USA in 2018 (cf. Roth 2018).

The attribution of a leading social function rubs off on the self-image of entrepreneurs. The global, mass use of their products elevates them, it seems, beyond the spheres of economics, even beyond those of politics. They are the all-powerful *Masters of the Universe*, influencing the lives of their users from Wall Street to Washington and far beyond. Sometimes these flights of fancy crystallize at individual events. Like Mark Zuckerberg's hearing before a committee of the European Parliament in the spring of 2018. It was about the data scandal surrounding Cambridge Analytica, which had caused a furor for months. After several difficult attempts to contact him, Zuckerberg accepted the invitation (not a summons) of the parliament. Then, during the committee, the surprise: no sign of a defensive stance in the face of the outraged public. Instead, self-confident maneuvering around the excited questions of the parliamentarians. In retrospect, it was measured: 62 min of questions,

23 min of answers (cf. Horn 2018). Zuckerberg seemed moderately impressed by the institution representing 500 million people, repeated the familiar and legally watertight phrases of his legal team and, after a good hour, pointed out that the scheduled time of the meeting had already been exceeded by 15 min. Zuckerberg passed over objections and follow-up questions with a stoic expression. The balance of power then became apparent at the very end of the meeting when the President of the Parliament, Tajani, himself ended a further discussion by referring to Zuckerberg's travel plans ("there is a flight"). A *Master of the Universe* answers what he wants for as long as he wants, even a President of Parliament knows that. 'My team will follow up' and 'I will send someone' were the winged words of a question-and-answer session that, instead of providing answers about Europeans' privacy, delivered a lesson in the power relationship between politics and the digital economy.

User numbers and economic and technological leadership seem to underpin the digital *masters'* understanding of power. Is this enough to legitimise power? Will this claim to leadership be fulfilled by technical achievements, will the promises of technology actually become reality? Anyone who has ever tried to get any information out of a voice assistant like Siri or Alexa apart from the weather or Wikipedia is likely to doubt this. Is the social pioneering role (including self-confident appearances in parliament) technologically justified?

The Cambridge Analytica scandal, at least, the reason Zuckerberg made a memorable appearance before the European Parliament, appears on closer inspection to be a storm in a teapot. The business model was as follows: Cambridge Analytica created a user profile based on a Facebook user's Likes (websites, movies, music, etc.), linked to a certain personality structure (distribution of points on five personality traits). Based on this analysis, users were shown ads that were supposed to be particularly effective for their personality structure. The method gained notoriety because Donald Trump used it for his pre-election campaign. Facebook banned third-party companies from analyzing users' Likes starting in 2015. So even before the 2016 presidential election. An influence of Cambridge Analytica on the election campaign itself is thus excluded. After the excitement had died down a bit, it also became clear that the reliability of reading off the personality structure on the basis of the Likes is very low. The mixture of flowery marketing statements by the CEO of Cambrige Analytica and the media excitement in connection with the Trump election led to the factually unfounded perception that Facebook users were being manipulated across the board and unknowingly in their voting decisions. These assumptions are not scientifically substantiated (cf. Rathi 2019).

Nevertheless, the trust in the digital infrastructure, in artificial intelligence and its predictions, seems great. After all, smartphones recognizes faces and talks to the user. Photo albums automatically recognizes who is in the holiday pictures. An artificial intelligence that reflects every analog movement of the user in the digital space. Is it just a matter of time before this intelligence takes on a life of its own? It doesn't seem to be that far yet. Shortcomings of technology are revealed, for example, in image recognition. Some AI systems cheat by using text elements of an image (e.g. copyrights) that allow conclusions to be drawn about the content. Or they can simply be misled by interspersed pixels that are invisible to the human eye (cf. Schmundt 2019). For voice assistants from Siri to Alexa and their dictation function, hundreds of (human) employees are employed to listen to the input of users and then transcribe it in order to improve speech recognition. This came to light through the complaints of these employees because they had to edit vulgar voice recordings (cf. dpa 2019). Anyone who whispers a flippant voice command into their phone should not think that the indiscretion will remain between them and their smartphone.

For the following investigation, the question arises: Are we now dealing with an omniscient digital infrastructure? The highlights of inadequacy do not fundamentally call into question the potential and ability of the digital infrastructure. They do not detract from its social significance. But they do show that behind the technical structures, there are people with interests—interests in money and power. And perhaps more than the current technical possibilities, it is the users themselves, their data and secrets, that help to enforce these interests.

1.4 Distinction from Cases Not Dealt with

In addition to the three levels we have examined—economic, social, and political—there are other theoretical building blocks and individual examples that, taken as a whole, lend momentum to the structural transformation of the private sphere. In the following, we list those that point beyond the theoretical framework of this book and are consequently not dealt with:

The Legal Level In the economic, political and social spheres, we are observing the structural transformation of the private sphere, without the legal status of privacy being significantly affected. In the medium term, however, it would also be the legal level that could develop a new understanding of privacy. In this case, however, we would no longer be dealing with a cultural or structural change, but with a shaking of the foundations of open and democratic societies. This is being prepared by a structural change in this area that is perceptible from weak indicators. (For more details: Afterword on the legal status of privacy at the end of the book by Bertil Cottier).

Privacy of Wolves in the Wild As the *Hello Barbie* example above points out, the digital age is also taking the privacy of those who are unaware of their private status to a new level. What applies to minors also applies to animals in the wild: recently, a mobile camera attached to a hunting dog in Sweden filmed an attack by two wolves on the same dog. And has gotten a lot of attention on YouTube. You can see two lone wolves roaming through the deserted forest. The two wolves interpreted the hunting dog as an intruder in their territory, whereupon it came to a bloody confrontation between dog and wolves. The hunting dog narrowly escaped. The video was posted online by the dog's owner and after just four days, the two lone wolves have already been viewed 2.6 million times. The wolf has no legal right to privacy. But the fact that the quintessential lone roamer is being followed by an entire online community points to profound changes in the private sphere.

Last retreats The structural transformation of the private sphere is powerful, but not absolute. In Dave Eggers' dystopian novel *The Circle*, it was the toilet where the protagonist found a last refuge of privacy for a few minutes. It almost seems as if these retreats are even harder to find today. This concern about latent surveillance was impressively expressed in the Oscar-winning documentary *Citizenfour* about the whistleblower Edward Snowden. In the presence of investigative journalists, Snowden, sitting on his hotel bed, has a red cloth handed to him, with which he covers himself and his laptop and only then agrees to enter his password. The fear of surveillance, of drones, sensors and cameras that could catch a high-resolution glimpse of the keyboard from outside, guided him to take this step. The blanket over his head as a last attempt to establish privacy.

Transparency and Ideology *For* the most part, a positive meaning is attributed to transparency, especially as a differentiation from manipulative concealment. Creating transparency means generating predictability and understanding through openness and information. This can be done as a volitional, individual process or as an external-investigative activity supra-individually (e.g. *Transparency (sic!) International*'s Perception Index, which measures corruption). Beyond the positive perception of transparency, there is, in the context of surveillance, a critical transparency that appears ideological:

Since the old public sphere is increasingly losing its function of controlling power and itself, i.e. of monitoring in an old-fashioned sense, the ideologists of the connecting and openness mission conclude that the private is also obsolete and the political. [...] The ideology of perfect transparency forms the correlate to the totalization of surveillance that is just taking place (Schneider 2015, p. 7).

We will discuss this ideological transparency and its political dimension, but in order to grasp the structural change of the private sphere, we will strive for neutrality with regard to a possible ideology of surveillance and transparency. In doing so, we rather stick to the concept of surveillance studies and 'liquid surveillance' (Bauman and Lyon 2013), i.e. the scientific observation and evaluation of surveillance.

To summarize the delimitations of our topic: For the theory of the structural change of the private sphere, and in order to theoretically develop privacy without secrecy, we limit ourselves to the privacy of individuals in the sense of subjects with legal capacity, elaborating the levels of economy, politics and social.

1.5   Does the Digital Age Still Need Theory?

The digital age is associated with a big claim regarding the legitimacy of theories: Big Data as the end of theory. At least the end of social science theories that model human behavior. Such models would become unnecessary because they could be comprehensively, realistically, and in some cases reproduced live on the basis of existing data (cf. Anderson 2008, p. 2). Looking at the data set tells it like it 'is', rendering theoretical superstructure unnecessary. Why then our *theory of the private without secrecy?*

Because the swan song for theory seems a bit overblown. Thanks to big data, the humanities and social sciences have elaborate possibilities to map reality with the new data. However, the flood of data cannot replace the fundamental function of theory, namely to provide an abstracted explanatory framework for observed phenomena. Without a theoretical superstructure, nothing can be discerned. Anyone who consults *Big Data* without theory only sees mountains of numbers. In order to *understand* the data volumes, theory is still needed.

In addition to the function of theory as an explanatory meta-construct, which is difficult to replace, it is above all the practical challenges that make the abolition of theory seem possibly rash, perhaps even completely exaggerated. Digital data are fragments and sections of complex realities of life. Their analysis faces the problem of the "messiness of reality" (Zwitter 2014, p. 2): does the totality of the collected (meta-)data form an accurate picture of what is called 'reality' in analogue perception? And how do image and reality relate to each other? As we can see, they are sometimes quite far apart. So far, in fact, that friendly people in the digital realm can turn into machos and rabble-rousers without this having any influence on their analogue selves—as we will describe in Sect. 4.3 on social structural change. Or to put it another way: the pleasure of playing a killer game does not produce a murderer (cf. Przybylski et al. 2014). Moreover, access to the fragmentary data is limited, as it is neither collected in one place nor collected by one entity. Business interests and, albeit eroding, confidentiality obligations do the rest. As a result, the informative value of the data and, consequently, the quality of the evaluation algorithms is highly variable.

Conversely, can these arguments also be applied to the secret private sphere in the digital world? All not so bad? Not at all: For the dissolution of the secret, it does not take the whole picture to bring about serious changes. Individual pieces of information brought to the outside world are enough, and in some circumstances even mere awareness of the possibility of this publication is enough. Unlike the researcher, who needs a comprehensive picture for description, informational highlights are enough for the bearers of personal data for political or economic use.

In short: The time of theory is not over. Especially for understanding the digital age and its impact on people, it is helpful. That is the reason for this book.

References

Anderson, Chris (2008): The End of Theory: The Data Deluge Makes the Scientific Method Obsolete. *Wired Magazine*. Abrufbar unter: http://archive.wired.com/science/discoveries/magazine/16-07/pb_theory *(letzter Zugriff: 22.10.2015)*.

Atwood, Margaret (2013): When Privacy is Theft. Review on The Circle by Dave Eggers. The New York Review of Books. http://www.nybooks.com/articles/2013/11/21/eggers-circle-when-privacy-is-theft/ *(letzter Zugriff: 22.04.2017)*.

Bauman, Zygmunt und Lyon, David (2013). *Liquid Surveillance: A Conversation*, Cambridge: Polity Press.

Bütler, Monika (2015): Abstimmen per Knopfdruck verändert die Entscheide. *NZZ am Sonntag*, Schweiz, October 4.

Cameron, David (2015): *The Guardian News*.

Dietz, Bernhard (2014): Diesseits und jenseits der Welt der Sozialwissenschaften. Zeitgeschichte als Geschichte normativer Konzepte und Konflikte in der Wirtschafts- und Arbeitswelt. In: Neuheiser, Jörg (Hg.): *Wertewandel in der Wirtschaft und Arbeitswelt Arbeit, Leistung und Führung in den 1970er und 1980er Jahren in der Bundesrepublik Deutschland*. Oldenburg: DeGruyter, 7–28.

dpa (2019): „Auch bei Facebook wurden Sprachaufnahmen abgetippt", Frankfurter Allgemeine, Diginomics, 14.08.

facebook (2014): Wie kann ich die Erkennung von Musik und Fernsehsendungen aktivieren oder deaktivieren? *facebook Hilfebereich*. Abrufbar unter: https://www.facebook.com/help/iphone-app/710615012295337 *(letzter Zugriff: 03.11.2015)*.

Floridi, Luciano (2014): *The fourth revolution: How the infosphere is reshaping human reality*. Oxford: Oxford University Press.

Garrett, Sean (2015): Early Twitter – Early Meerkat. *Twitter.com*. Abrufbar unter: https://twitter.com/SG/status/574755572542652416 *(letzter Zugriff: 16.07.2015)*.

Heffernan, Virginia (2016): *Magic and loss: the Internet as art*. New York: Simon and Schuster.

Henkel, Christiane Hanna (2019): „Google und Tesla statt General Electric und GM: wie Amerikas Firmenlandschaft innerhalb eines Jahrzehnts umgekrempelt wurde", NZZ, Wirtschaft, 5. August.

Homer; Ebener, Dietrich (1976): *Homer: Werke*. Berlin: Aufbau-Verlag.

Horn, Dennis (2018): *Peinlich für die EU*. tagesschau.de. https://www.tagesschau.de/ausland/zuckerberg-vor-eu-ausschuss-101.html (06.08.2019).

Kroker, Michael (2015a): Instagram lässt Twitter weit hinter sich – und die facebook-Familie ist 2,8 Milliarden Nutzer groß. *WirtschaftsWoche-Blog*. Abrufbar unter: http://blog.wiwo.de/look-at-it/2015/09/29/instagram-lasst-twitter-weit-hinter-sich-und-facebook-familie-ist-28-milliarden-nutzer-gros/ *(letzter Zugriff: 19.04.2016)*.

Kroker, Michael (2015b): User nutzen Social Media fast 1,8 Stunden am Tag – 30 Prozent der gesamten Internet-Zeit. *WirtschaftsWoche-Blog*. Abrufbar unter: http://blog.wiwo.de/look-at-it/2015/10/20/user-nutzen-social-media-fast-18-stunden-am-tag-30-prozent-der-gesamten-internet-zeit/ *(letzter Zugriff: 19.04.2016)*.

Lanier, Jaron (2014): *Who owns the future?* New York: Simon and Schuster.

McLuhan, Marshall; Fiore, Quentin; Agel, Jerome (2005 [1967]): *The Medium is the Massage*. Corte Madera: Gingko Press.

Neumann, Linus (2015): BigBrotherAward 2015 für „Hello Barbie". *BigBrotherAwards.de*. Abrufbar unter: https://bigbrotherawards.de/2015/technik-hello-barbie *(letzter Zugriff: 03.08.2015)*.

Precht, Richard David (2007): *Wer bin ich – und wenn ja, wie viele? Eine philosophische Reise*. München: Goldmann Verlag.

Przybylski, Andrew K; Deci, Edward L; Deci, Edward; et al. (2014): Competence-impeding electronic games and players' aggressive feelings, thoughts, and behaviors. *Journal of Personality and Social Psychology* 106 (3), S. 441–457.

Rathi, Rahul (2019): *Effect of Cambridge Analytica's Facebook ads on the 2016 US Presidential Election*. Medium/Towards Data Science. https://towardsdatascience.com/effect-of-cambridge-analyticas-facebook-ads-on-the-2016-us-presidential-election-dacb5462155d (06.08.2019).

Ribi, Thomas (2016): Das Flüstern der Dinge. *NZZ Feuilleton*. Abrufbar unter: http://www.nzz.ch/feuilleton/chancen-der-digitalisierung-das-fluestern-der-dinge-ld.16058 *(letzter Zugriff: 26.04.2016).*

Roth, Daniel (2018): *LinkedIn Top Companies 2018: Where the U.S. wants to work now.* LinkedIn Pulse. https://www.linkedin.com/pulse/linkedin-top-companies-2018-where-us-wants-work-now-daniel-roth/ (13.08.19).

Schmidt, Eric (2010): *The Washington Ideas Forum.*

Schmundt, Hilmar (2019): „Wenn Computer mogeln", DER SPIEGEL, Wissen, 09.08.2019.

Schneider, Manfred (2015): Ende des Gesellschaftsvertrages, Aufstieg der Überwachungskultur. *NZZ-Podium vom 24. September 2015 «Überwachungskultur».* Abrufbar unter: http://podium.nzz.ch/event/uberwachungskultur/ *(letzter Zugriff: 29.09.2015).*

Spehr, Michael (2015): Ausgespäht mit Android. *Frankfurter Allgemeine,* Computer & Internet, August 6.

Tryfonas, T., Carter, M., Crick, T. and Andriotis, P. (2016): Mass Surveillance in Cyberspace and the Lost Art of Keeping a Secret – Policy Lessons for Government After the Snowden Leaks. Human Aspects of Information Security, Privacy, and Trust. Heidelberg: Springer. 174–185.

Werner, Ella Carina (2015): Vom Fachmann für Kenner. *Titanic-Magazin.* Abrufbar unter: http://www.titanic-magazin.de/fachmann/ *(letzter Zugriff: 11.12.2015).*

Zwitter, Andrej (2014): Big Data ethics. *Big Data & Society* (July–December), S. 1–6.

Part I

The Secret Private: Introduction and Derivation

2

"Privacy Is Dead": How Could It Come to This?

Since Habermas we recognize the disintegration of the clear dividing lines in industrial society between state and society, between private and public as the "structural transformation of the public sphere" (Habermas 1990). We elaborate this theory as "structural change of the private" for the digital age. Our basic thesis is that the private in the digital age no longer knows any secrets. So how do privacy and secrecy relate to each other in the face of the disruptive changes brought about by the digitalization of all areas of life? We answer this question along a transformation of omniscience and its changes:

- After religion, as the bearer of a transcendent omniscience, had given way to the Enlightenment, there was a short phase of *immanent* knowledge, i.e. knowledge related to the worldly-experiential, and thus the possibility of *individual secrecy*. We call this possibility, in demarcation to (divine) omniscience, *analogous self-knowledge*.
- Now, however, the introduction of the digital into all aspects of life leads to *digital, omniscience,* which is, however, *immanent*. Secrecy comes under pressure from this omniscience. We just haven't fully realized it yet.

The transition from *self-knowledge* to *omniscience* is the common thread in the following case studies and in the later development of the theory of a *privacy without secrecy*. First, however, in the following section we ask: How could it come to this? Under what circumstances was it possible for a privacy without secrecy to develop?

The translation of this chapter was done with the help of artificial intelligence (machine translation by the service DeepL.com). A subsequent human revision was done primarily in terms of content.

To address these questions, we first bring the *secret* into the discussion of the *private*. We situate ourselves in the discourse around these two concepts, which are central for us. To this end, we introduce the individual functions of *the private*, and subsequently three forms of the *secret* within this private Sect. 2.1. We focus on the epochal threshold that has brought us to the current situation (cf. Seele 2008, 2018): from religious, *transcendent analogue omniscience* via *analogue intrinsic knowledge* to *immanent digital omniscience*. We then outline the social preconditions that are necessary for digital omniscience Sect. 2.2: Mass society and its drive towards counterculture. Within this economic-social framework, we identify the *Rebel Sell* as a door opener for the private without secrets.

We begin with the description of the first basic concept: What does the *private* mean, from which the secret is said to have separated? What form and function does it have for the individual?

2.1 What Is the Private Sphere?

What is *private* can first be described in contrast to what is *public:* Private is what is not public. The one defines itself in distinction from the other. In the case of privacy, this has been a tradition since ancient Greece: on the one hand, the *polis,* the public life of the citizen. On the other hand, the sphere of the *oikos,* the household, which is separated from the polis and over which each individual looks out for himself (cf. Habermas 1990, p. 56). To this day, social analysts rely on this dichotomy, with which the "basic structure of these societies can be grasped" (Ritter 2008, p. 9).

This definition is not entirely without presupposition. The pair of opposites *private–public* is associated with the implicit thesis that the two actually represent something separate. This separation is not uncontroversial in contemporary research. Rather, the two concepts are perceived as dynamic and merging levels in the reality of life (cf. Klaus 2001, p. 24 f.). However, for our concern to describe the *private* via its functions for the individual, the analytical separation is helpful. Habermas clarifies this approach by drawing a dividing line along the political function:

> The public sphere is limited to public violence. (Habermas 1990, p. 90)

It is only through the possibility of not participating in this public sphere that the separability of the two spheres is constituted. This is also indicated by the Latin root of the term: *privatus* (adj.) which means, among other things, *without public office*. This brings us closer to the second half of Habermas's definition:

> The private sphere also includes the actual 'public sphere'; for it is a public sphere of private people. Within the realm reserved for private individuals, we therefore distinguish privacy and the public sphere. (ibid.)

This private public sphere, constituted by aggregated private individuals, is to be distinguished from the state-institutional "public power". Our investigation, however, is concerned with the individual level, which Habermas calls the private sphere within the private sphere:

> The private sphere encompasses bourgeois society in the narrower sense, i.e. the sphere of commodity transactions and social labour; the family with its intimate sphere is embedded in it. (ibid.)

The described *intimate* sphere within bourgeois society is the area that is turned away from the public sphere, which is of particular interest to us within Habermas' conception of privacy. Here it is not so much the framework of the family, but the individual member and his or her intimacy in the sense of his or her inwardness, his or her world of feelings and thoughts, coupled with the acting out of love, affection and caring for oneself and others. The private sphere—let us include Habermas' "private sphere" and "private sphere" under this term—is the precondition for the intimate sphere and determines its extent and form. The private creates the framework in which intimacy emerges and protects it from access by others (cf. DeCew 2015, p. 9). For Habermas, this private sphere is—or, in view of digitalisation, should be said: was—the protective cloak of privacy.

In Habermas' conception, the person with his or her privacy is endowed with *private autonomy*. This concept further delimits our object of investigation. It connects the aspect of privacy with the state of autonomy. If, as previously stated, the private is thought of as distinct from public authority, the economy, and the family, an immature, dependent person has a hard time asserting his or her private as protected and secret.[1] The autonomous person, on the other hand, is responsible for his own private sphere.

---

[1] It should be emphasized that Habermas sees precisely the economy (and not politics) as the midwife and guarantor of private autonomy: "For to the extent of the enforcement of the capitalist mode of production promoted from above, social relations are mediated by relations of exchange. With the expansion and release of this sphere of the market, the owners of commodities gain private autonomy; the positive sense of 'private' is formed in the first place at the concept of the free disposal of property functioning capitalistically" (Habermas 1990, p. 143).

Quite in this sense, for Hannah Arendt, the *private* sphere is the sphere in which the free person fulfils his or her necessities of life, the sphere of one's own free household and housekeeping (cf. Arendt 1998). These private affairs, insofar as they are secured and covered, enable the actor to pass into the public sphere and to step out of privacy. The free, private person as a precondition for the free public person (cf. Jacobitti 1991; Morariu 2011).

Every free individual deals with his privacy differently—and decides for himself when to step out of privacy. The private sphere thus takes on the character of a volitional, inner mode. This is how it comes up in an essay describing the current change in the world of work:

> Place or time no longer play a role for work and private life. Work is done in the café, cooking in the kitchen—the change between the private sphere and the world of work only takes place in the mind. At the kitchen table or in the café, one dives into work or back into private life. (Böttcher 2013, p. 95)

In distinguishing it from work, the immaterial, individual character of privacy becomes clear. It is not a material space or a specific time that determines the private sphere, but a *decision*. *What is* decisive is the possibility of being able to choose between the modalities and to have the certainty that the private sphere will be preserved if necessary. The *private* is thus linked to the free, autonomous human being, to a person who has decision-making power, who can actively decide for or against sharing something of his private (cf. Lanier 2010, p. 4). Hereby we approach the notion of *informational self-determination*, which is a central function of the private. Having clarified the social location of privacy, we approach it further through its functions for the individual.

Here, too, we begin *ex negativo* and open with a demarcation: the private as the assured absence of judgments and measurements about the individual by third parties and the inaccessibility of informational bases for such assessments (cf. Introna 1997, p. 263). The definition revolves around the notions of judging and measuring. The private prevents these assessments from intruding into the intimate sphere. The aspect of *assured absence* is important. It is the apprehension of judgment or condemnation by others not taken into confidence that feeds the desire for the private. The apprehension is sufficient here—an actual invasion need not occur at all to formulate such a claim to privacy.

The private sphere is dependent on its social circumstances—on time and space. A look at the sleeping habits of days gone by shows how relatively briefly a secure, self-evident private sphere existed. As late as the beginning of the nineteenth century, it was common for several people—in inns even strangers—to share a bed. Today the bed stands symbolically for a place where one enjoys an intimate sphere protected by privacy, shared with a partner of one's own choosing (cf. W 2015). This change is complemented by a cultural boundedness of privacy. This can be well observed in the perception of *personal space*—the space that a person internally perceives as their own. Whereas in North-Western societies 'an arm's length' is regarded as the required distance and falling short of it leads to resentment, in other cultures—the image of a subway filled to bursting point in Japan or a train in India comes to mind—this appears to be of lesser importance (cf. DeCew 2015, p. 12).

In summary, we consider the *private* sphere as a claim of persons endowed with private autonomy, outside of politics and social work within the framework of a bourgeois society. The function of *the private* is the secured absence of judgements and condemnations by third parties and thus the protection of an intimate sphere. The private stands as a screen in front of this intimate sphere and makes it possible in the first place. An intimate sphere without the private is not possible.[2]

Individual parts of the intimate sphere contain that *secret* with which we deal in greater detail below: Aspects of the thought world and associated emotions that do not want to be communicated to the outside world. If the secret is communicated after all, this is linked to great trust.

So how can we imagine this secret that we encounter protected by the private and shrouded in privacy?

2.2 The Secret Private: Three Types

According to Derrida, the secret is always linked to a negation—it is something that either may not be said at all or may not be said further.[3] It is through negation that the secret acquires its character (cf. Derrida 1992; Lawlor 2014, p. 7). The secret is a piece of information whose public disclosure is associated by the bearer of the secret with consequences that are undesirable for him. The secret we are interested in presupposes a person and his social embeddedness. The information declared secret would have a certain effect within this social embedding. This description implies that shaping a piece of information into a secret is a conscious act. However, in view of our

---

[2] The other way round, as we will see in the course of further investigation, this is not true: a private without an intimate exists.
[3] According to Derrida, institutions and individuals can be bearers of secrets. In the following, we will deal with the secrets of individuals. State and corporate secrets are also affected by digitalization. We deal with the effects of digitalization on human life and leave out Area 51 and the recipe of Coke.

thesis that the secret has departed from the private, this understanding expands. It still includes the consciously 'produced' secret: a diary entry, a confidential chat, a personal photograph. But the term must be supplemented by that secret of which one is not necessarily aware, but the disclosure of which would nevertheless have a serious impact on privacy in the sense of the definition above. Metadata is such a case: where one is and when one is with whom, this information is in principle secret in the case of a private errand. But one is not even aware of a secret at the moment of the transaction—the movement, the trip, the meeting with others.

We identify three types of a *secret* private at the level of the individual in the following sections, which we define as follows (Fig. 2.1):

1. *Transcendent analogous omniscience* is, under the assumption of an omniscient divine authority, the awareness of the constant co-knowledge of one's own secret private 'from above'. From this, the standardization of one's own behavior follows. Like the famous booklet of Peter before the gates of heaven, in which all good and bad deeds are recorded in bookkeeping to a moral life balance.
2. *Immanent analogous 'self-knowledge'* is the awareness of an individual secret that no neighbor, judge or God knows, unless one consciously chooses to share this secret private. The folk song *Die Gedanken sind frei (Thoughts are Free)* vividly sums up this hermetic understanding of *the* secret private: "*I think what I want/and what makes me happy/but everything in silence,/and as it is fitting./My desire and wish/no one can deny,/it remains so:/the thoughts are free*".
3. *Immanent digital omniscience* is the awareness of the possibility that through digital infrastructures everything secret private can be accessed at any time and from anywhere. As with type 1, this *immanent digital omniscience has a normative effect on one's own behaviour*: The individual subjects himself

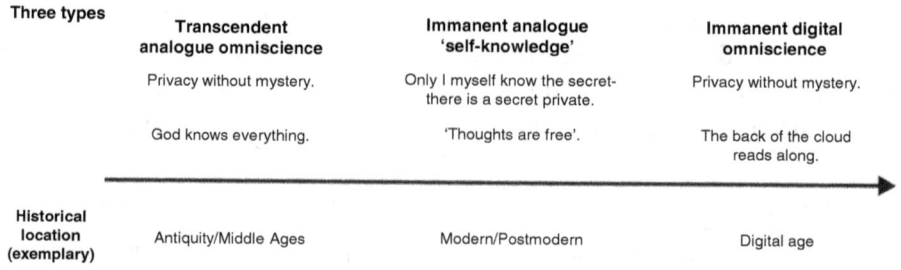

Fig. 2.1 Is there a secret private and who knows about it? Three types

to self-censorship without any direct external effect, a 'scissors in the head' is established.

For the following theory of the structural change of the private, type 3 in particular, the *immanent digital omniscience* with regard to the secret private, is of importance. With case studies and theory building, we crystallize the transitions between the three types of the secret private and describe the new quality of the private in the digital age. Although the transition from analog to digital infrastructures suggests a temporal ordering of these three types, they are not to be understood strictly diachronically. Rather, all three types occur synchronously in the digital age.[4] Accordingly, there are different forms of the secret private at the same time.

As a first step, we present the three types in detail, examining the relationship between privacy and secrecy and how it is evolving.

2.2.1 God Sees Everything: Transcendent Analogous Omniscience

The first part of our typology of the secret private describes the *transcendent analogous omniscience*. By the existence of *omniscience*, a private *secret* accessible only to the individual is excluded. Thus, in this type, there are no aspects of the world of thought excluded from joint knowledge. Informational self-determination is suspended vis-à-vis the omniscient entity. Control by a God or Gods takes on a behavior-channeling effect that enabled cooperation in the first place, but also triggered conflict as in the concept of BigGods (cf. Norenzayan 2013 for religion in general, as a Big Data science fiction novel building on the ideas of Dirk Helbing's iGod of Dicke 2017 and Zapf 2015 in relation to Protestantism). For the believer, there is an awareness that a secret from other people is not secret from the god or gods.

This omniscience is further characterized by two clarifications:

- Omniscience is *transcendent*. This transcendence refers to the frame of reference of *omniscience*—not a worldly entity, but an extra-worldly entity endowed with omniscience, one of the classical predicates of God. *Transcendent* here means 'other-worldly', standing above the empirically perceptible and the social.

---

[4] On the other hand, type 3 is bound to digital infrastructures and accordingly, in a diachronic perspective, cannot be found historically with types 1 and 2.

- Omniscience is *analog*. With regard to our questioning of the secret private in the digital age, this *analogue* means the non-availability of digital infrastructures as carriers and collectors of this omniscience.

The transcendent frame of reference already hints at it: This form of the non-secret private in the analogue refers primarily to the subject of religion. By way of illustration, let us consider the Christian manifestation, in which an omnipotent God symbolizes transcendent omniscience. A God who, according to tradition, is above all things human and all social processes. This omniscience does not need digital channels. For its dissemination, faith and, as a concretization, the printed Holy Scriptures are sufficient.

The transcendent analogous omniscience of the Christian God is elaborated in several places in the Bible, especially in the Old Testament. For example in the first Book of Samuel, where it says[5]:

> But the LORD said unto Samuel, Look not on his appearance, and on his tall stature; I have rejected him. For the LORD looketh not on that which a man looketh on. A man looketh on that which is before his eyes: but the LORD looketh on the heart. (1 Sam 16:7)

While the human observer must limit himself to the perception of external appearances, God looks directly into the heart of man—and thus also sees all that remains hidden from human observation in private, secret. In view of God's creation of man, this ability seems obvious: "He who planted the ear, should he not hear? He who made the eye, should he not see?" (Psalter 94:9). As Creator, He has full access to the sensory faculties of His creatures. And profoundly so, as becomes clear a few lines further on: "But the Lord knows the thoughts of men: they are but a breath!" (Psalter 94:11).

A few chapters further on, in Psalter 139:1–4, it says:

> A psalm of David, to be sung. Lord, thou searchest me, and knowest me. I sit or rise, thou knowest; thou understandest my thoughts afar off. I walk or lie down, thou art about me, and seest all my ways. For, behold, there is not a word upon my tongue, Which thou, O Lord, knowest not already.

Not even the smallest movement remains hidden from God. Both physically ('sitting or standing') and spiritually: even from a distance, God recognizes

---

[5] The following Bible excerpts are all from Luther's Einheitsübersetzung, 1984 edition.

people's thoughts and even when they are not expressed. And it does not stop with the mere observation of the secret. God also processes the secret further:

> But there is a God in heaven who can reveal secrets. He made known to King Nebuchadnezzar what was to happen in the days to come. Your dream and your visions, when you were asleep, were like this [...]. (Daniel 2:28)

The Christian believing individual—at least if Bible-bound—must assume in view of such an image of God that his secrets are observed and then evaluated by a transcendent entity. Even unspoken and secret thoughts and plans are revealed to God. In the internal religious perspective, this *secret private* is relevant to faith. The omniscient entity becomes the overseer of godly living. God knows, and God knows what is right and to be done, and God also knows who does it right. Thus, theological concepts become binding in the inner life as well. God's secular agents in parsonages and churches function in this system as sharers of transcendent omniscience. The believer entrusts himself to them in this conviction and confesses to them face to face, thus making God's worldly agent on earth a confidant and judge of the secret private.

Awareness of the non-existent secret can work in two directions for the believing individual. Divine omniscience can be associated with positive effects: an increase in generosity, a higher level of cooperation, less deceptive behavior—even among strangers. Divine oversight reduces conflict (see Norenzayan 2013, p. 23). However, transcendent omniscience can also have a negative effect, for example causing a latent guilty conscience. For man fails, be it in himself or in the religious demands made of him. Consequently, one's way of life is no longer adapted out of insight, but out of fear of surveillance and the inner and outer world consequences.

This omniscience-induced fear is now met by the Enlightenment. And clears away the divine omniscience, freeing one from transcendent observation and immaturity. Man is thrown back upon himself: Now only oneself is all knowing and we become our own inner-worldly judge. God is removed from the equation.

2.2.2 "Secret Privates": Immanent Analogical Self-Knowledge

With the removal of transcendent omniscience from consciousness, self-knowledge becomes possible. With far-reaching consequences: Suddenly, there is no automatic co-knowledge. Instead, there is a secret private.

Historically, this is to be seen as a short phase, and during its existence as something new.

What exactly is meant by this second type of the secret private, the *immanent analogous omniscience*?

- *Self-knowledge* is central. The secret private lies solely with the individual. There is no unwilling or unwitting complicity. Self-knowledge is limited only by individual non-awareness.
- *Immanent* is this self-knowledge because it is not questioned or undermined by any supernatural power. This power is abolished. The knowledge of one's own secret private is bound to the inner-worldly person (and its finiteness).
- Self-knowledge is *analogue* because it exists in the analogue, inner-worldly privacy of the individual. It is disconnected from a digital, networked infrastructure. It is offline.

There is no universal conception of a secret private sphere and its significance that applies to all people throughout the world: self-knowledge is shaped by a particular culture, philosophy and political attitudes, and the demand for and protection of the secret private sphere depends on social developments. It is thus relative and shaped by a collective consciousness. We already mentioned the Enlightenment in this context, but also more recent events such as the totalitarian regimes of the twentieth century, all-encompassing spying such as that of the GDR State Security or the revelation of the wiretapping methods of US and British secret services shape an awareness of the urgency of a secret private (cf. e.g. Jarvis 2011, p. 42 f.). The secured secret stands in opposition to systems that approach the individual in an encroaching manner. This is also the view of the German Federal Constitutional Court in its 1983 ruling, widely perceived as a milestone of data protection. In the context of data collection for a planned census and the resistance in providing a wide variety of information to officials going from door to door, the constitutional judges clarify:

> Whoever cannot foresee with sufficient certainty what information concerning him is known in certain areas of his social environment, and whoever is not to some extent able to estimate the knowledge of possible communication partners, can be substantially inhibited in his freedom to plan or decide on the basis of his own self-determination [...] Whoever is uncertain whether deviating behaviour will be noted at any time and permanently stored, used or passed on as information, will try not to attract attention by such behaviour [and] possibly

forego an exercise of his corresponding fundamental rights [...]. This would not only impair the individual's chances of development, but also the common good, because self-determination is an elementary functional condition of a free democratic community based on the ability of its citizens to act and participate. (Bundesverfassungsgericht 1983, Section II, C)

The constitutional judges state: The free democratic state is essentially based on the secret private of its citizens. Citizens must not be harassed by any force in its secrecy, must be able to develop their own attitudes and convictions undisturbed by influences of power, which enable them to fight the battle of opinions in a lively, pluralistic democracy. The private and the secret are the individual spaces which show the state its limits and at the same time make it possible in its liberal expression.

But the secret private can also be a space in the most literal sense. As Virginia Woolf, the British writer and advocate for the women's movement, describes it in her 1929 essay: *A Room of One's Own* (Woolf 1977). A room of one's own, a space for undisturbed creative activity. The possibility of closing a door, this was for Woolf the precondition of emancipation and for women to be as intellectually productive and brilliant as men.

But the secret, and with it the immanent, analogous self-knowledge, is not limited to one's own four walls. One can also be completely secret and private outside the home, for example when strolling with no known destination, on a Sunday walk, with only one's own head and one's own thoughts, without having to answer to anyone. Without a name tag, without a navigation system. Self-knowledge as an inner space, perhaps even more important than material space, a state: psychic privacy with complete informational self-determination. The assured knowledge that there is no complicity.

In this spirit, liberal constitutional states grant their citizens personal rights and the protection of privacy. This creates a secure space in which the citizen can develop. The data on the person and his home are explicitly protected. And with it their self-knowledge. State and non-state encroachments into this space are sanctioned (see Schmale and Tinnefeld 2014, p. 116). Informational self-determination of one's self-knowledge also includes selective openness to third parties. If the secret private is shared, communication about it becomes a secret matter. For this reason, the channels of communication are also specially protected by law. The secrecy of correspondence and the secrecy of telecommunications guarantee the citizen that his or her shared private knowledge will only be shared with a third party. Only in exceptional cases—for example, for prison inmates—this right is justified and temporarily suspended.

It is precisely this example of the secrecy of communication that makes clear the attacks to which the secret private sphere is exposed: eavesdropping attacks erode self-knowledge. As soon as religion has been relativized as a threat to self-knowledge, the next actors blow up a storm on the secret private sphere: rulers, secret services, and corporations push so close to the individual that he or she loses his or her immanent, analogous self-knowledge bit by bit. Here, once again, the historical short-term nature of actual self-knowledge, the historical exceptional situation of complete availability over one's own information, becomes clear: It was limited to a short period after the Enlightenment and before digitalisation. In the digital age, the secret private no longer exists.

2.2.3 "Privacy Without Secrecy": Immanent Digital Omniscience

The digital age is characterized by the omnipresence and constant availability of information (cf. Helbing 2015, p. 104)—and the more information is produced, according to our thesis to be proven in the following, the less there is a secret private sphere. This connection was already apparent in the analogue, in the gigantic surveillance apparatuses of the twentieth century: Gestapo, KGB, CIA, state security—admittedly with a different ideological backgrounds but functionally similar—were the precursors of institutions such as the NSA or Google and Facebook, which today are comprehensively informed about the lives of their fellow human beings with digital support and use this information according to their own ideas.

The secret private of the free citizen is no longer inviolable (cf. Ball 2016, p. 6): Where surveillance is anchored in consciousness, it leads to self-censorship. Where it goes unnoticed, it leads to the loss of informational self-determination. Type 3, the *immanent digital omniscience* is therefore characterized by

- After a short but formative phase of self-knowledge, now there is *omniscience* again. Co-knowers who have access to the data traces of everyday life. Almost all traces. The result is a complete, comprehensive picture of the individual, of which the individual may have no idea.
- *The digital* is the new quality of omniscience. The pervasive use of digital infrastructures that permeate all aspects of life and thus the emergence of a digital self is the basis of immanent omniscience. Digital media are tapped

as carriers of the secret private—texts, images, voice recordings. Digital sensors monitor and record the behavior of users.

- The omniscient is brought to earth and thus *immanent*. Through digital technology. Thus the potential omniscience is used all too humanly and on this side, exploited politically and economically. In the divine case, omniscience was believed to be transcendent: although with influence on the inner-worldly-human sphere, it was removed from it. Not bound by its laws, transcending its nullities and limitations. In the case of digital omniscience, the knowing entity is very concrete on earth, immanent, even materially locatable, even if rather abstractly present in everyday life: As a server in huge data centers, as a visible and visitable website, as a company, as a logo, as concrete information.

Digital omniscience replaces analog intrinsic knowledge. Although it is immanent, there are also connections in this new omniscience to that transcendent omniscience that seemed marginalized in modern, enlightened societies. Networked activities are perceived as *omnipresent, omniscient* and *omnipotent* (see Helbing 2015, p. 81 for an account in relation to Google). Whether superhumanly perceived or not, the secret private is severely affected by the digital—it changes behaviours. In everyday offline life, one's privacy is unconsciously and incessantly considered. An automatic privacy filter: you guard your secret, you don't share all the information available to you with all your conversation partners. Let alone letting passers-by or strangers in on it, you choose your confidants carefully. Depending on the interlocutor, information is filtered, weighted, perhaps even glossed over or omitted. Communication takes place selectively with targeted addressees. If the communication is secret, it is whispered. Secret private information that is not to be shared with others remains within one's own reach, for example in a locked diary or only as a thought (Fig. 2.2).

The digital equivalent of the private sphere, on the other hand, presents itself differently, it is more difficult to grasp than its counterpart in the analogue world. It has no clear contours. Digital offerings that contain personal

Fig. 2.2 Analogous places of the secret private

information—e.g. a digital diary stored in the cloud, emails, chat and video telephony, etc.—add additional intermediaries to the storage and communication of the secret private in the analogue comparison. There are added the terminal devices involved as well as the digital storage locations. First of all, one's own terminal device, computer, mobile phone, tablet, becomes the immediate confidant of the secret private, and in the case of communication, additionally the terminal device of the recipient. In the case of online storage, the storage location ("the cloud") becomes the further carrier of the secrets. In addition, there is a grey area of secret carriers: unnoticed collected and stored metadata such as the location and usage behavior, which are stored in the user profile (Fig. 2.3).

In this new cartography, the locations of the secret private have multiplied. Moreover, in addition to one's own consciousness, one's own terminal device is now the most direct bearer of the secret private. The extent to which this has been internalised is shown by the queasy feeling when a stranger handles one's own mobile phone or computer and thus transgresses the secret private in the digital analogue. The uneasiness when insight is taken into the place of origin of the digital self: Browser history, text and image documents, music, contacts. Who was last written to, what was searched for online, which pages were saved, what was bought, whose messages were ignored. This digital privacy is distinct from analog access by fellow human beings. From family and friends, who should not know everything. From strangers, should they find or steal

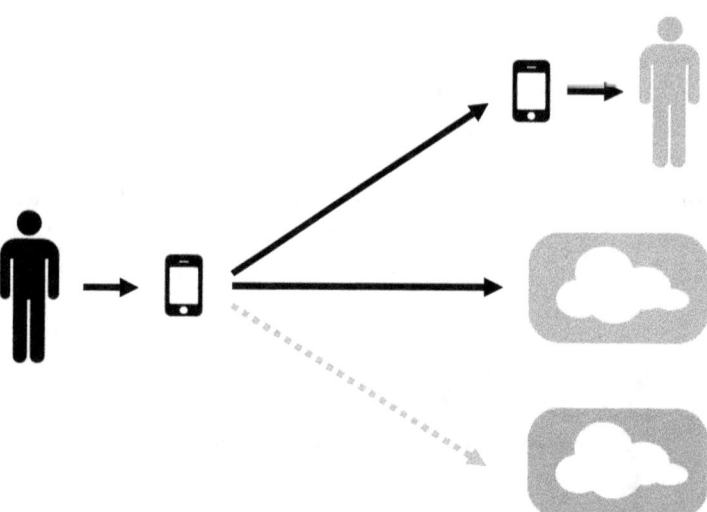

Fig. 2.3 Digital places of the secret private

the device because they could do harm with it: Identity theft, blackmail, fear of being vulnerable to the digital secrets on the device.

This digital privacy is central to trust in the connected digital infrastructures and the corresponding products. It's why Facebook regularly reminds its users to change their passwords. It's why Apple is outspoken in refusing to build a backdoor into the encryption of its products, and it is why Google is dedicated to fighting phishing. The secret private should be protected from unauthorized access in its various digital locations. Because the data that is supposed to be protected from the foreign gaze has long since been processed, evaluated and used. Invisibly used, in fact, by the very companies that just reminded us to change our passwords regularly.

The secret private in the digital is an incomplete privacy: the personal environment, the real tangible fellow human beings may not directly learn of the secrets. But a server somewhere remembers everything.

At least on the corporate side, no secret is made of the data collection. What is logged and for what purpose is—according to the law—publicly accessible (cf. Google 2015). Only the results of the gigantic data collection remain secret. The secret of the private individual seems to be worse protected than the results of the data collectors. Access to the collected data is thereby reserved for a few companies and state institutions. Thus, the individual, private secrets outside the individual suddenly become economic and political secrets, "[…] troves of dossiers on the private lives and inner beings of ordinary people, collected over digital networks, […] packaged into a new private form of elite" (Lanier 2014, p. 127).

The digital collection of secrets undermines the secret private sphere as an achievement and legal asset (cf. Schmale and Tinnefeld 2014, p. 120 f.). An individual encryption of this data could reopen the path to informational self-determination, contribute to the establishment of a digital *secret* private. A digital whisper. But the methods are under political fire and are suspected across the board of aiding terrorism and other criminal activities (cf. e.g. McAfee 2016, p. 4; Reuters 2016). Without considering the proportionality between benefits and potential abuse. Politicians want to keep a loophole open into the secret private sphere of their citizens, to secure digital omniscience for themselves in the future.

The immanent digital omniscience cares little about the political rights of the citizen and the criticism of control. It follows its own logic. The logic of digital omniscience is based on the asymmetry of information between the observer and the observed. It is pre-emptive, does not need a specific occasion and does not need to be on the spot. And it relies only selectively on human intervention, because it is automated. It needs no reason, because it is *always-on*. Digital omniscience is post-discursive and post-ideological, because it relies on automated algorithms that recognize only mathematical sequences instead of content and controversy. Ideology and discussion take place on a different plane. The logic of digital omniscience is purely goal-oriented. It knows no difference between private and public, between political, social and economic. It is a de-differentiating logic (cf. Andrejevic 2016, who elaborated on these attributes in relation to drones, as well as Hillenbrand's science fiction novel "Dronenland" (2014) on the subject).

As this logic permeates parts of our everyday life, we have arrived in an age of all-encompassing availability of personal information (cf. Floridi 2014, p. 219), experiencing in this respect a "time-ontological transformation" (Seele 2016a, b). An instant-transparency that leads us into the *post-privacy society*. This is the society Dave Eggers introduces us to in his dystopian novel *The Circle* (2013). In his world, the slogan *Privacy is* theft! To keep personal information secret from the public is to withhold something from the public. Privacy becomes subject to justification This reversal of values need not be a problem. It can also be seen, here and now, in best management fashion, as an opportunity: New "life management strategies" emerge from the abolition of privacy (Heller 2011). The abolition of the secret private leads to transparency that goes so far as to exclude unfair behaviour by individuals and at the same time prevents the misuse of this information (cf. Introna 1997, p. 260). This ethic of unconditional transparency is the appropriate ethic for the "infosphere", a living space under the auspices of all-encompassing information with the goal of peaceful coexistence. Jeff Jarvis, a tech blogger from the United States, illustrates this with a simple example:

> When Steve tells Bob that he is getting divorced, Bob is the one facing decisions about what to do with that knowledge. Bob should ascertain whether he has Steve's permission to tell others. Bob should ask himself why he'd pass it on—to gossip and hurt Steve or to gather support around Steve and help him. (Jarvis 2011, p. 131)

In this perspective, the actual bearer of the secret—Steve—is no longer perceived as an actor at all. It is up to the new bearer of the secret-private information—Bob—to do something useful with it: Not blaspheme, not gossip, but help. The world as a utopian garden, driven by complete information, transparency and charity.

Theories such as these legitimise the abolition of the secret private sphere and ascribe to it an overriding benefit. Complete transparency as a vision has existed for a long time, crystallized in the "panopticon". The philosopher and pioneer of utilitarianism, Jeremy Bentham (1748–1832), is regarded as the creator and author of this theory. Since then, panopticon theories have also been used to describe digitalisation (for an overview of the digital panopticon, synopticon or post-panopticon, see Seele 2016a, b). The radical mind conceived of an architecture that would allow for the most efficient surveillance possible and architecturally revolutionize institutions with a need for supervision—e.g. prisons, schools, and industrial plants. The goal was to expose inmates to constant visibility. Complete transparency aimed at creating conformity (see Bentham and Bozovic 1995). The inmates of the panopticon are thus to be observed from a central location, usually a tower. They cannot see the surveillant because the inside of the surveillance tower remains dark. The surveilled never know exactly when or if they are being watched. This leads to behavioural adaptation, from within the individual, based on ignorance (cf. MacCannell 2011, p. 29). Building on this effect, it is thus architecture alone that leads to the normalization of behaviors (cf. Garland 2001). This architectural idea drew wide circles. For example, at the centre of the first Great Exhibition in London in 1851 was the *Crystal Palace*, which with its glass walls was designed in such a way that visitors could potentially be observed at all times. The museum order was thus architecturally guaranteed. Such standardization can also be observed in other contexts: The *Hawthorne effect*, named after a study of labor productivity in which it was found that workers showed an increase in performance only because they were being watched. In crime prevention, *Anticipatory Effects* are observed when robberies decrease due to knowledge of video surveillance (see Smith et al. 2002). For everyday civilian life, there is the concept of *self-censorship*, by which people react to a presumed observation voluntarily and without external influence by adapting their behavior. For example, no pointed political opinions are expressed in an online forum anymore if it is assumed that the entries are being monitored and the statements could lead to suspicions of extremism. Sanctions are feared. Although neither is being monitored, nor is the author an extremist, nor is the behavior even worthy of sanction.

Michel Foucault extended this idea from an architectural concept to a principle of order in modern Western societies. According to this, various instances of surveillance and control lead to the disciplining of citizens, above all he names submission to the functioning of the market as a symbol of this disciplining. The panopticon as a power technique (cf. Foucault and Seitter 1977).

In the digital age, the form of the panopticon is changing. It is no longer an architectural, place- and time-bound form of exercising power. Digital omniscience is a liquefied panopticon that penetrates closer to people than ever before, independent of time and place. Every participant, every *user* eagerly and voluntarily builds into it through their usage behavior (cf. Seele 2016a, b; Wewer 2013, p. 70). The digital panopticon operates with key concepts such as transparency, networking, *sharing of* knowledge and skills. The digital panopticon is invisible, impalpable, naturally painless. It is uncertain if or when it observes and it increases the effect of self-control and self-censorship.

In the place of the darkened Benthamian watchtower, there is the possibility that somewhere a tapped fiber optic cable might be recording, recording, evaluating digital information. The immanent, digital omniscience functions as an elaborate panopticon. And is correspondingly successful in standardizing behavior.

2.2.4 Encroachments on the Secret Private in Digital Omniscience

The immanent digital omniscience has a serious impact on the secret private lives of people whose personal information is part of the digital infrastructure. It is not electronic data processing per se or 'Big Data' that is central here. It is not 'the' electronic media that concern us here. It is specific aspects of digital information collection and processing that can be traced back to individuals—and that are concerned with the secret private sphere of these individuals.

Within the digital infrastructure, the secret private is initially just one piece of information among many others, collected, stored and evaluated in countless applications. On the one hand, impersonal information is used for statistical purposes for science, governmental or corporate tasks without repercussions for the originator, to understand contexts, to prepare 'soft' or 'hard' legislation (Cominetti and Seele 2016) or to conduct market research. On the other hand, digital infrastructure also allows the collection and processing of individually traceable data—along with personal internet accesses and unique user data this is technically easy. Compared to impersonal information, individual traceability expands the purposes for which it can be used: for government use, opportunities arise for security and law enforcement agencies to confront and control the information originator. For businesses, it becomes possible to target customers personally and promote individually tailored products. Based on data knowledge, preliminary decisions can be made about the originator of the data. For example, by pre-selecting certain advertised products or, more drastically and in terms of political consequences, by increasing personal control or surveillance. Maturity is reduced. Moreover, and through the awareness of surveillance, the use of secret-private,

2 "Privacy Is Dead": How Could It Come to This?

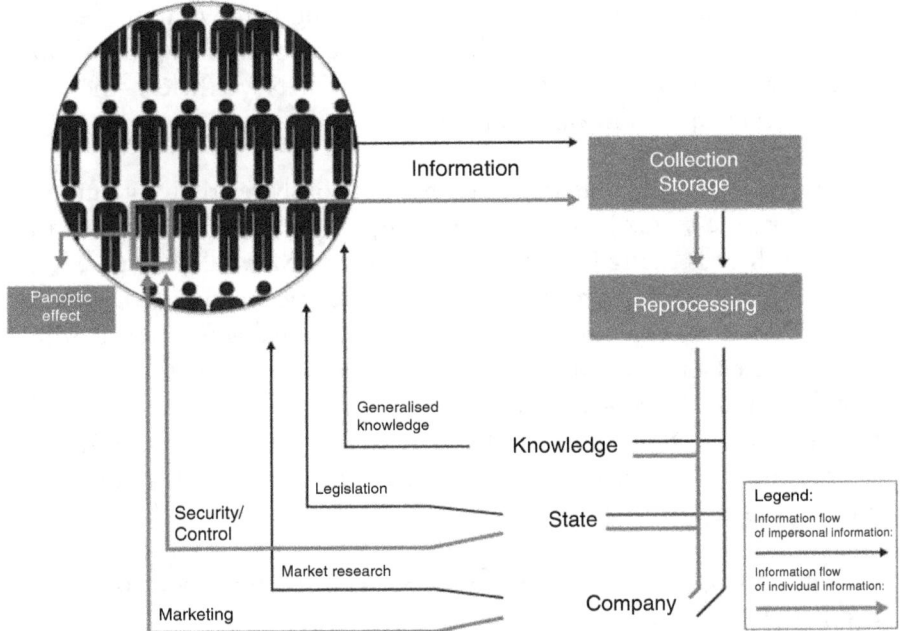

Fig. 2.4 Encroachment on the individual in digital information processing

individualized data has an impact on its creator: the behavior of the monitored adapts—willingly or unwillingly. The digital panopticon has an effect.

This problem—and thus the precise subject of our representations—is illustrated in the following diagram (Fig. 2.4).

The diagram shows the assaults to which the individual and his secret private are exposed by digital information processing. In the following section, we supplement this individual perspective with the accompanying changes that occur on the societal level that arise with the immanent, digital omniscience.

2.3 Mass Society as a Social Precondition of Immanent-Digital Omniscience

How were the infrastructures responsible for the radical transformation of privacy and secrecy in the digital able to establish themselves so quickly, voluntarily and comprehensively? The social context seems central to this question. So in what follows, we switch the level of observation from the individual to society and embed the development in a grand theory of social structural change: The theory of mass society. We are dealing here with an ideal type with which changes in society as a whole are to be traced. The model assumptions should be viewed with appropriate caution. Nevertheless, they offer clear analytical added value on two levels:

1. The aspect of mass *expansion of* offerings previously restrictively limited to a wealthy elite seems helpful, because digitization and its impact on the private sphere is doing just that, in the spirit of Marshall McLuhan's dictum: "Print technology created the public. Electric technology created the mass" (McLuhan et al. 2005, p. 68). Just as the analogue and the printed media created a public, the digital created offerings for the masses.
2. The finding of the *economic primacy of* the mass society seems interesting for our field. Entrepreneurial rather than governmental activities are decisive for the establishment of digital omniscience, as we will see in the second section of this chapter.

Accordingly, we update the theory of mass society with the aspects of the digital and therein the secret.

2.3.1 Historical Materialism and De-individualization

Mass society is based on free civil society and the accompanying economic changes: The social shackles of feudalism and class society are cast off. Through the abolition of serfdom and the free choice of employment, broader social strata gain economic relevance. Economic development is the result: a civil society emerges in which it is not supposedly divinely foreseen rulers who determine one's social position, but one's own economic fortune, or more precisely: one's economic success. Social change is based on the changed mode of production, society can accordingly be described as an economically dominated power structure, even if this rests on a democratic-legal foundation as development progresses. The theory of mass society thus takes a materialist understanding of history as its starting point (cf. Labriola 1966).

The theory describes society after industrialization and democratization. According to Habermas, the defining characteristics of mass society are *expansion* and *participation*. Power, education, merit and culture, previously the preserve of a small elite, are now accessible to a broad mass: mass media, mass democracy, mass culture, mass production. According to Habermas, every aspect of expanded participation is thereby oriented towards its economic function—a "commercial re-functioning" in all areas of society, market events are altogether leading the way (Habermas 1990, p. 257). The consequence of this economic orientation can be seen in culture, more precisely: in mass culture. It is now no longer oriented towards an absolute culture, but only towards a relative one, since it is market-mediated:

> It is no longer merely the mediation and selection, presentation and decoration of works—but their production as such that is oriented in the broad areas of consumer culture according to aspects of sales strategy. Yes, mass culture acquires its dubious name precisely by the fact that its expanded turnover is achieved by adapting to the relaxation and entertainment needs of consumer groups with a relatively low standard of education, instead of conversely educating the expanded audience to a culture intact in its substance. (Habermas 1990, p. 254)

The critical theory of the Frankfurt School resonates here.[6] Mass culture as the enemy of everything "that would be more than itself" (Horkheimer and Adorno 2000, p. 165). Due to the appealing and universal suitability for the masses and the form of the market, cultural decay occurs, in the sense of a decline in quality, borne and accelerated by the mass adoption of this culture. A *tyranny of the masses* prevails, which always perpetuates and consolidates that which it dominates, namely the producers and the existing relations of production. The latter, coupled with industrialization, increasing mechanization and rationalization, have brought about a steadily rising overall social prosperity and thus a growing sales market. With rising prosperity, the possibilities for consumption increase, and the boosted consumption goes hand in hand with standardization—this enables mass production, depresses prices, and incidentally shapes the perception of consumers:

> As the industrial machinery produced standardised goods, so did the psychology of consumerisation attempt to forge a notion of the 'mass' as 'practically identical in all mental and social characteristics'. Consumerism must therefore be a system of rigid conformity. (Heath and Potter 2005, p. 110)

The equality of consumer goods, according to the thesis, leads to the equality of the consumer. Thus, mass society achieves totality: standardization of

---

[6] Here, special reference should be made to the context in which Habermas's research originated, Germany in the 1960s. One recognizes a conception of culture that is, if not elitist, at least substantialist. Culture "intact in its substance" (Habermas 1990, p. 254) is, one would argue, merely a bourgeois culture. The fact that high literature and opera belong to this category may be traditionally justifiable, but it does not seem self-evident as absolute 'substance'—especially in contrast to the critically considered mass culture. 'Good' culture, for Habermas, is the culture of education and aspiration. Accordingly, the author calls for a bourgeois cultural education of the masses instead of the "lowering" (ibid.) of culture to the level of the masses.

people according to the requirements of mass production. The participants are absorbed comprehensively: Everyone is simultaneously employee, profiteer and consumer (cf. Bell 1956). The economy is not limited to economic activity. It takes on a normative form that encompasses the whole of life and becomes the primacy of society:

> The attitude that everyone is forced to adopt in order to prove their moral suitability for this society over and over again is reminiscent of those boys who, when admitted to the tribe, move around in circles smiling stereotypically under the priest's blows. Existence in late capitalism is a constant rite of passage. Everyone must show that he identifies without remainder with the power by which he is beaten. (Horkheimer and Adorno 2000, p. 187)

Coercion, moral aptitude, beatings, constant smiling. This is how critical theory portrays mass society, admittedly in a grand tradition: we know capitalism not only as an economic, but also as a cultural and social form from Marx, and the associated perception as a history of decay as well. Minus the polemics, the claims of exposure and change, and the historical conditionality of this theory, it remains to be said: Mass society has an effect on the people living in it and from it through mass, industrial production.

For the establishment of immanent digital omniscience, we take away two insights: The standardizing character of the system causes most members of society to be directly or indirectly involved in economic processes. At the same time, market processes have penetrated all aspects of life and can be influenced accordingly by (large-scale) entrepreneurial products and strategies. Market in full social breadth and depth.

The stereotypical image of mass society is the grey army of office workers. Hat and trench coat, on their way to work, a uniformly serving and consuming mass, supported and made unquestionable by economic prosperity and stability. In this dystopian ideal type of mass society, there is no deviation from the norm, from the standard—the mass is determinative, the individual follows. The time period of this society is the heyday of industrialization in the nineteenth century and the following extended prosperity until the middle of the twentieth century (cf. Schneider 2001).

Nor does this diagnosis fit with the picture we know: Apple, Google, Twitter seem to represent the exact opposite of this absolute, capitalist standardization. They present themselves as individualistic, self-determined, different, hip. Not standardized. We resolve this contradiction in the following section.

2.3.2 The Drive towards Individualisation: Counterculture, *Rebel Sell* and Digitalisation

In the middle of the twentieth century, a deviation from mass society is becoming apparent. The coercive, normative character of this society awakens unpleasant memories of the polarized masses of the Third Reich. The images of the frenetic, militarized people of a Nuremberg party congress and subsequent terror and destruction contributed to the fact that

> W]hat had previously been only a moderate distaste for conformity, common among artists and romantics, got pumped up unto a hypertrophied abhorrence of anything that ever smacked of regulatory or predictability. Conformity was elevated to the status of a cardinal sin, and mass society became the dominant image of a modern dystopia. (Heath and Potter 2005, p. 327)

The will to distance oneself from the uniformity and standardization of mass society is, of course, not exclusively connected to the experience of aggressive, National Socialist standardization. But this creates a basis for opposition to standardization and the totalitarian features of this social meta-order. Equally clear is the positioning against communism as it was pursued in the West after the end of the Second World War: The hunt for communist ideas bears obdurate features in the USA and Europe of the 1950s; think of the 'attitude check' for civil servants in Germany, the Fichen affair in Switzerland, or the political discrimination against artists and politicians that went down in the history books as *McCarthyism*. Democracy, full shelves in the supermarket and a new car against egalitarianism and Soviet uniformity.

On top of the distinction from mass society stands the growing importance of individualisation as a diffuse umbrella concept (cf. Berger 2010). According to Sennett, the self comes to the fore at the expense of the mass-controlled public sphere and contributes to its destruction (cf. Sennett 2003; Schneider 2001). A counterculture emerges that confronts mass society with its own ideas. The main culture in this picture is still the industrial-market economy mass society. The counterculture deviates from this, goes into opposition to standardisation and central regulation, and offers an alternative that takes the individual as its starting point: Consumer ethics instead of prohibitions (Schmidt and Seele 2012), self-help instead of prescribed programs, individual spirituality instead of state church, reform pedagogy instead of state schools (cf. Heath and Potter 2005, p. 327 f.).

The focus on the individual influences the economic ideas within the counterculture. One is not turned away from the economy, but would like to give it a more personal touch. This is not possible in mass society, because production and consumption take place impersonally, en masse. The counterculture stands against this, it wants to offer products that are individually tailored, decentralized. The goal is not standardization, but the distinction of the consumer. Consumption is linked to the representation of status and lifestyle, a rebellion against the established mainstream (cf. Heath and Potter 2005, p. 330). The *Rebel Sell* combines the need for individuality, for rejection of the mainstream and standardization, the desire for uniqueness and the pleasure and participation in consumption. The act of buying as a social, political, economic and ecological statement at the same time.

The economic potential of this new mode of consumption is being given a huge boost by the digital revolution, represented here by the computer industry and the *New Economy*: more products, more co-determination, decentralised structures. With social and technical development, the economic order of mass society is changing.[7]

We are now in the early 1980s. Computers are no longer a foreign word, but digitization is not yet a mass phenomenon; mainframes dominate the business. Computer companies—above all IBM—are large and cumbersome, entrusted with dry subjects such as accounting and administration, which are associated with the mass collection of data and its processing. Computers are economic and political instruments. Into this scepticism, the PC enters—the *personal* computer. No longer as a distant antagonist, but as a personal and private companion (cf. Lotter 2013, p. 23). The computer steps out of the grey administrative building into the progress-oriented public. The early home computer user sets himself apart from the crowd. It is offbeat, rebellious and different (cf. Wasserman 2005). Paradigmatic of this change is a 1983 Apple TV commercial. The title and aesthetics of the one-minute spot called *1984* are reminiscent of the Orwell film adaptation of the same name: a grey mass of uniformly dressed workers stares at a giant screen. A Big Brother-like figure delivers a speech, touting the *Unification of Thoughts* as an achievement, *one will, one people, one cause* as the goal. A good-looking female athlete, unlike her uniformed, will-o'-the-wisp surroundings, in colorful clothing, sprints toward the giant screen lulling the masses and destroys it by hurling a huge hammer at it. The text fades in: "On January 24th, Apple Computer will introduce Macintosh. And you'll see why 1984 won't be like '1984'" (Hayden

---

[7] We already encounter the connection between technology, economy, social change and power in Marx's interpretation of industrialization and the increasing importance of capital, and it is also echoed in critical theory: "What is concealed here is that the ground on which technology gains power over society is the power of the economically strongest over society. Technical rationality today is the rationality of domination itself" (Horkheimer and Adorno 2000, p. 147).

2011). What is being advertised here is not a computer—but an individual, well-designed, emancipative product. The rejection of mass-societal uniformity, uniformity and remote control become the sales argument, the new product the salvation from grey uniformity. At the presentation of the commercial, Steve Jobs, then CEO, clarifies Apple's role in this emancipative process: it is about standing up to IBM, the biggest competitor at the time, and putting its monopoly into perspective, with a better product and a better superstructure: the buyers of the Macintosh should perceive Apple "as the only force that can ensure their future freedom"—after the sentence, cheers break out in the audience. Apple sells much more than a computer with its product. It is a counter-cultural statement, individual, it fights against the mass grey, does things differently. It positions manufacturers and buyers in opposition to the standardized mainstream. Always on behalf of the very big goals: Creativity, emancipation, freedom. A self-image to which the company remains true 30 years later and under new leadership: in 2016, Apple publicly refused to help the FBI decrypt an iPhone used by a terrorist. Tim Cook, now CEO of the company, addressed all customers with the following lines:

> While we believe the FBI's intentions are good, it would be wrong for the government to force us to build a backdoor into our products. And ultimately, we fear that this demand would undermine the very freedoms and liberty our government is meant to protect. (Cook 2016)

If the US administration itself doesn't do it, the company will have to uphold the most American of all virtues itself: Freedom and Liberty. Apple serves a self-created, counter-cultural niche—computer pioneers, against the establishment, big industry, standardization. From the very beginning, the company has targeted creatives, intellectuals, dissidents and those who would like to be. It is no coincidence that "Think different" was the company slogan from 1997 to 2002. An overall product that did not lose its countercultural flair even after it had long since arrived in the mainstream.

With the home computer, its inventors and manufacturers, turtlenecks, jeans and sneakers move from nerdy tech revolutionaries to the middle of society. Steve Jobs has arrived. And with all the force of a new industry equipped with global demand. The digital as the hinge of the counterculture into mass culture.

The case of Apple exemplifies that the counterculture has no reservations about making money. Contrary to what the left-dominated, critical initial thinking might have suggested, the scepticism towards the economy refers primarily to large corporations and their claim to power and standardisation.

People *do* not want to be dictated to by this kind of capital, but want to realize their own projects—*do it yourself!* as the rallying cry of a popularized revolt against the system, which has become so broadly anchored that it has itself become the system (cf. Rapp 2008). Following this insight, Slavoj Žižek calls the actors of this revolt *liberal communists:* smart, dynamic, nomadic, fighting against established structures and authorities. Flexible, with culture and education against routine, against the values of mass society and the standardized *look and feel of* industrial production. The old camps of right and left no longer fit for these actors: *liberal communists* are successful managers, especially in Silicon Valley (cf. Žižek 2006). They are precisely those managers who, digitally supported, abolish the secret private sphere—always relaxed, without a suit and tie. Wolves in hoodie fur.

We see a counterculture-inspired industry that has arrived in the mainstream and the mass market: the counterculture makes peace with the masses (cf. Heath and Potter 2005, p. 328). Newly added to mass society are the attributes *digitized* and *individualized.* The countercultural movement has not redeemed its potential for change, but rather turns out to be a mass society with an updated face. This is particularly evident in the loose, start-up flair, individuality and freedom promising agents of an industry whose stock market determined corporate values now far exceed those of the classical industries with their plants, smokestacks and standardized armies of workers in front. In the hip look of the *New Economy,* the old structures are hiding, new elites are forming in similarly closed circles as before (cf. Andrejevic 2004, p. 55).

A breeding ground for the abolition of the secret private in the digital age.

2.3.3 Summary: Digital, Individualised Mass Society and the Abolition of the Secret Private Sphere

So, in summary, how did the digital infrastructures that did away with the secret private get established so quickly, voluntarily, and across the board?

The starting point for the considerations was an early form of mass society, characterized by mass production, mass culture and standardization. A primacy of economy prevails, which takes hold of society in its full breadth and depth. With increasing prosperity, the need for a non-conformist, more individual counter-culture develops. This counterculture, above it settles into the economic cycle, becomes linked to business interests. It leads to a form of mass society updated by the *digital* and *the individual.*

And with momentous implications for the secret private. From the beginning of its dissemination in the administrations of politics and business, computer technology was criticized as a means of surveillance. The systematic collection of personal data, activities, work results, residence, religion, everything that could suddenly be processed digitally en masse, was perceived as an encroachment on the private sphere (cf. Lotter 2013). The symbol of the dystopian data collectors was IBM, and the defining political event was the German census of 1983: *Dataization! Computer state! Information expropriation!* were the slogans used by critics (cf. Berlinghoff 2013, p. 14). The ruling of the Federal Constitutional Court against the census and for the strengthening of informational self-determination represents the culmination of the early revolt against the attack on the secret private sphere. At the same time, the mass distribution of home computers began. Unlike the mainframe computer in the administration building, the home computer is very close to the user. A digital counter-design to a peeping administration. Apple cleverly positioned itself in this market, as described above using the example of the Orwell commercial: A counterpoint to the data-collecting threats to freedom. The Macintosh as a statement against large corporations and centralist surveillance,[8] a device very close to the digital private needs of the user (cf. Zittrain 2011; Farber 2014).

The history of digitalisation shows the necessity of relating the structural change of the private sphere to the change of the underlying media. Let us look at digital communication technologies, accordingly, the area in which digitalisation has the most obvious practical effects on life. Marshall McLuhan and his famous saying: 'The Medium is the Massage!' comes to mind here.[9] According to McLuhan, the extent of the changes associated with the establishment of new media cannot be overestimated. All spheres of life are profoundly affected by it, the new media massaging change through their use. What McLuhan cites for "electrical technology" applies equally to the social impact of digital infrastructures:

---

[8] Apple could not initially win its lead and the ideological as well as technological battle against IBM. After a weakness in the home computer sector in the early 1980s, IBM had caught on over the course of the decade and developed its own offerings for office and home. In conjunction with Microsoft operating systems, the concept prevailed over its competitors—namely Commodore, Amiga and Apple. The basis for nationwide, individual digitalisation was thus laid—without Apple. This consolidated Apple's countercultural image, with a positive influence on market shares since the late 1990s.

[9] Mind you, not message, but *massage*. What initially began as a typo in the first edition was adopted by McLuhan in terms of content. He wanted to counter the cliché of his statement and at the same time illustrate the massaging character of the new media (at that time television) (cf. McLuhan 2016).

> The medium, or process, of our time—electric technology—is reshaping and restructuring patterns of social interdependence and every aspect of our personal life. It is forcing us to reconsider and re-evaluate practically every thought, every action, and every institution formerly taken for granted. Everything is changing—you, your family, your neighborhood, your education, your job, your government, your relation to 'the others'. And they're changing dramatically. Societies have always been shaped more by the nature of the media by which men communicate than by the content of the communication. (McLuhan et al. 2005, p. 8)

Media are people's window into the world and therefore significantly shape what people perceive through this window (cf. Carr 2011, p. 17). In the case of digital communication technologies, this shaping is accompanied by a high degree of self-purposefulness of the medium: the technical structure is both the starting point and the objective of action—it is the means of communication and its purpose. The constant confrontation with products that are supposed to solve an individual 'problem' create their own set of problems and solutions. The products create their own demand and establish a self-sufficient logic of their own that is adopted by the individuals using them. In technology studies, this logic is referred to as *persuasive computing*. Devices expand the capabilities of their users, provide them with experiences, and establish new relationships. Computers increase the efficiency of their users, reduce barriers by providing more information, facilitate decision making, and thereby carry the user experience close to the individual and his or her private. In this process, the digital infrastructure, initially only a means to achieve the set goal, itself becomes a factor influencing the goal (cf. Fogg 1998, p. 227). In its pronounced form, this mixing of means and ends is referred to as 'solutionism', a primacy of the technical solution over the actual understanding of the problem. Due to real-time digital confrontation, people position themselves on all kinds of issues "before they have understood them" (Kaube 2013).

With the progressive use of digital infrastructures, the nature and consequently the content of communication is changing: they are aligned with the technical possibilities. Content is supposed to be faster to receive, more up-to-date and more tailored to individual interests (cf. Grabowicz 2014). The digital transforms communicative one-way streets into interactions embedded in a multitude of other information: From where is communication taking place, how, for how long, and with what? In this way, in addition to the information itself, a great deal of further information, some of it personal, is transported. This requires a digital infrastructure, knowledge about the users and their individual preferences. Thus, as digital home systems become more widespread, thanks to digital information collection, insights are gained for further differentiation of the offer. In return for the disclosure of personal data, the digital economy promises the perfectly fitting, tailored product. The consumer is supposedly addressed personally, with his or her individual needs. In this context, the individual with his or her preferences for the creation of content plays a minor role: what counts are aggregated patterns of action that are extracted from the data stream flowing to the companies and that become the sales argument (Andrejevic 2016, p. 35).[10]

The following graph represents the argument described in the previous section (Fig. 2.5):

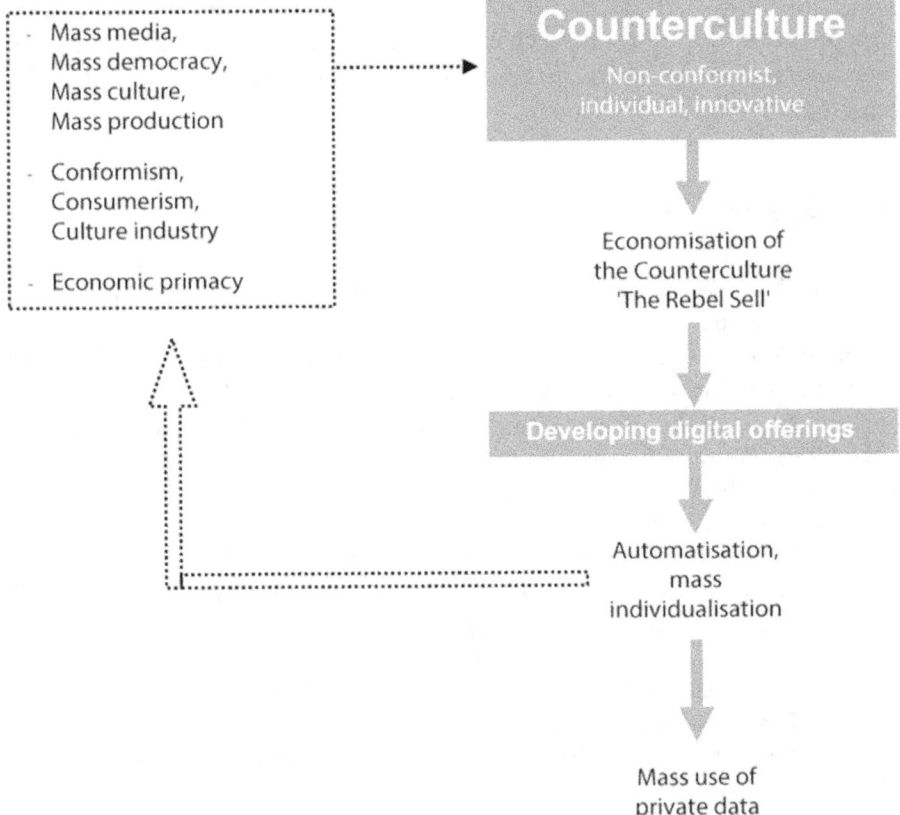

Fig. 2.5 Social preconditions of mass society

---

[10] There are many examples of this kind of individualization: Deutsche Post offers individual design options for envelopes, an online doll store lets customers choose their hair color and hairstyle, Amazon customizes its home page according to the "browsing trends" of other customers, and thanks to the numerous configuration options and features, virtually no newly produced car that rolls off the assembly line is exactly identical to another. The decisive factor in these individualizations, however, are not the wishes of the individual, but the anticipation of these wishes based on knowledge of collective customer's wishes.

The private data are, along the lowest arrow in the graph, voluntarily revealed by the users due to a high subjective plausibility. The promise: individual satisfaction of needs through products that are close to the user. Entering personal data is easy if the information serves to improve and tailor the offer. In this way, the digitalised, individualised mass society and its major economic players act as a door-opener to the abolition of the secret private sphere at the individual level.

In the following sections, we apply the preliminary considerations on the structural change of the private sphere to lifeworld examples. Subsequently, we explain the consequences theoretically. We illustrate the abolition of the secret private and the structural change of the private with a leading question: How can the private be thought without the secret—in economy, society and politics?

References

Andrejevic, Mark (2004): *Reality TV: The work of being watched.* Lanham: Rowman & Littlefield Publishers.

Andrejevic, Mark (2016): Theorizing Drones and Droning Theory. In: Ders. (Hg.): *Drones and Unmanned Aerial Systems.* Heidelberg: Springer, S. 21–43.

Arendt, Hannah (1998 [1958]): *The human condition.* 2. Aufl. Chicago: University of Chicago Press.

Ball, Kirstie (2016): All consuming surveillance: surveillance as marketplace icon. *Consumption Markets & Culture* S. 1–6.

Bell, Daniel (1956): The Theory of Mass Society. *Commentary* 22 (1), S. 75–88.

Bentham, Jeremy; Bozovic, Miran (1995): *The Panopticon Wriitngs.* London: Verso.

Berger, Peter (2010): Alte und neue Wege der Individualisierung. In: Berger, Peter (Hg.): *Individualisierungen.* Wiesbaden: VS-Verlag, S. 11–25.

Berlinghoff, Marcel (2013): Computerisierung und Privatheit – Historische Perspektiven. *Aus Politik und Zeitgeschichte* 15–16 S. 14–19.

Böttcher, Dirk (2013): Das Private? Ist in Arbeit. *Brand Eins* 08/13 (Schwerpunkt: Privat), S. 90–95.

Bundesverfassungsgericht (1983): *Volkszählungsurteil.* Karlsruhe: Urkundsstelle des BVerfG.

Carr, Nicholas (2011): *The shallows: What the Internet is doing to our brains.* New York, London: WW Norton & Company.

Cominetti, Marta und Seele, Peter (2016). Hard soft law or soft hard law? A content analysis of CSR guidelines typologized along hybrid legal status. *uwf UmweltWirtschaftsForum* 24 (2–3), 127–140.

Cook, Tim (2016): A Message to Our Customers. *apple.com*. Abrufbar unter: http://www.apple.com/customer-letter/ *(letzter Zugriff: 18.02.2016).*
DeCew, Judith (2015): Privacy. *Stanford Encyclopedia of Philosophy.* Abrufbar unter: http://plato.stanford.edu/entries/privacy/ *(letzter Zugriff: 02.03.2016).*
Derrida, Jacques (1992): How to Avoid Speaking: Denials. In: Coward, Harold; Foshay, Toby (Hg.): *Derrida and Negative Theology.* Albany: SUNY Press, S. 73–142.
Dicke, Willemijn (2017): *iGod.* Amazon: CreateSpace Independent Publishing Platform.
Farber, Dan (2014): The Macintosh turns 30: Going the distance. *CNET.* Abrufbar unter: http://www.cnet.com/news/the-macintosh-turns-30-going-the-distance/ *(letzter Zugriff: 07.04.2016).*
Floridi, Luciano (2014): *The fourth revolution: How the infosphere is reshaping human reality.* Oxford: Oxford University Press.
Fogg, Brian J (1998): Persuasive computers: perspectives and research directions. *Proceedings of the SIGCHI conference on Human factors in computing systems* 98 S. 225–232.
Foucault, Michel; Seitter, Walter (1977): *Überwachen und Strafen: Die Geburt des Gefängnisses.* Frankfurt a.M.: Suhrkamp.
Garland, David (2001): *The Culture of Control: Crime and Social Order in late Modernity.* Chicago: The University of Chicago Press.
Google (2015): Hinweise zum Datenschutz bei Google. *Google-Account Information.* Abrufbar unter: https://accounts.google.com/signin/privacyreminder/c?continue=https%3A%2F%2Fdrive.google.com%2F%23#cbstate=12 *(letzter Zugriff: 08.09.2015).*
Grabowicz, Paul (2014): Tutorial: The Transition To Digital Journalism. *Berkeley Advanced Media Institute.* Abrufbar unter: https://multimedia.journalism.berkeley.edu/tutorials/digital-transform/ *(letzter Zugriff: 16.06.2016).*
Habermas, Jürgen (1990 [1962]): *Strukturwandel der Öffentlichkeit: Untersuchungen zu einer Kategorie der bürgerlichen Gesellschaft.* Frankfurt am Main: Suhrkamp.
Hayden, Steve (2011): ‚1984': As Good as It Gets. *Adweek.* Abrufbar unter: http://www.adweek.com/news/advertising-branding/1984-good-it-gets-125608 *(letzter Zugriff: 20.01.2016).*
Heath, Joseph; Potter, Andrew (2005): *The Rebel Sell. Why the Culture can't be jammed.* Chichester: Capstone.
Helbing, Dirk (2015): *The Automation of Society is Next How to Survive the Digital Revolution.* Zürich: ResearchGate.
Heller, Christian (2011): *Post Privacy: Prima Leben ohne Privatsphäre.* München: Beck.
Horkheimer, Max; Adorno, Theodor W (2000): *Dialektik der Aufklärung: Philosophische Fragmente.* Frankfurt am Main: Fischer Taschenbuch Verlag.
Introna, Lucas (1997): Privacy and the computer: why we need privacy in the information society. *Metaphilosophy* 28 (3), S. 259–275.

Jacobitti, Suzanne Duvall (1991): The public, the private, the moral: Hannah Arendt and political morality. *International Political Science Review* 12 (4), S. 281–293.

Jarvis, Jeff (2011): *Public parts: How sharing in the digital age improves the way we work and live.* New York: Simon and Schuster.

Kaube, Jürgen (2013): Ist Ingenieur sein denn glamourös? *Frankfurter Allgemeine*, Feuilleton, December 9.

Klaus, Elisabeth (2001): Das Öffentliche im Privaten – Das Private im Öffentlichen. In: Klaus, Elisabeth: *Tabubruch als Programm.* Opladen: Leske + Budrich, S. 15–35.

Labriola, Antonio (1966 [1903]): *Essays on the Materialistic Conception of History.* New York: Monthly Review Press.

Lanier, Jaron (2010): *You are not a gadget. A Manifesto.* New York: Knopf Press.

Lanier, Jaron (2014): *Who owns the future?* New York: Simon and Schuster.

Lawlor, Leonard (2014): Jacques Derrida. *Stanford Encyclopedia of Philosophy.* Abrufbar unter: http://plato.stanford.edu/entries/derrida/ *(letzter Zugriff: 02.03.2016).*

Lotter, Wolf (2013): Die Ruhestörung. *Brand Eins* 08/13 (Schwerpunkt: Privat), S. 22–27.

MacCannell, Dean (2011): *The ethics of sightseeing.* Berkeley: University of California Press.

McAfee, John (2016): The Obama administration doesn't understand what ‚privacy' means – let me explain. *Business Insider.* Abrufbar unter: http://www.businessinsider.de/john-mcafee-obama-administration-privacy-2016-1?r=US&IR=T *(letzter Zugriff: 25.01.2016).*

McLuhan, Eric (2016): Commonly Asked Questions (and Answers). *The Marshall McLuhan Estate.* Abrufbar unter: http://www.marshallmcluhan.com/common-questions/ *(letzter Zugriff: 21.06.2016).*

McLuhan, Marshall; Fiore, Quentin; Agel, Jerome (2005 [1967]): *The Medium is the Massage.* Corte Madera: Gingko Press.

Morariu, Mihaela (2011): Public and Private in the Anthropology of Hannah Arendt. *Agathos: An International Review of the Humanities and Social Sciences* 2 (2/2011), S. 146–150.

Norenzayan, Ara (2013): *Big gods: How Religion Transformed Cooperation and Conflict.* Princeton: Princeton University Press.

Rapp, Tobias (2008): Plädoyer gegen Punks: Die dümmste Jugendkultur. *taz*, Gesellschaft/Alltag, November 14.

Reuters (2016): Obama wirbt für Zugriff auf Handys in Ausnahmefällen. *Reuters.* Abrufbar unter: http://de.reuters.com/article/usa-apple-obama-idDEKCN0WE0NO *(letzter Zugriff: 13.03.2016).*

Ritter, Martina (2008): *Die Dynamik von Privatheit und Öffentlichkeit in modernen Gesellschaften.* Wiesbaden: Springer.

Schmale, Wolfgang; Tinnefeld, Marie-Theres (2014): *Privatheit im digitalen Zeitalter.* Wien: Böhlau.

Schmidt, Imke; Seele, Peter (2012): Konsumentenverantwortung in der Wirtschaftsethik: Ein Beitrag aus Sicht der Lebensstilforschung. In: *Zeitschrift für Wirtschafts- und Unternehmensethik*. 13/2. 169–191.

Schneider, Irmela (2001): Theorien des Intimen und Privaten. In: Schneider, Irmela: *Tabubruch als Programm*. Opladen: Leske + Budrich, S. 37–48.

Seele, Peter (2008): *Philosophie der Epochenschwelle: Augustin zwischen Antike und Mittelalter*. Berlin und New York: Walter de Gruyter.

Seele, Peter (2016a): Digitally unified reporting: how XBRL-based real-time transparency helps in combining integrated sustainability reporting and performance control. *Journal of Cleaner Production*.

Seele, Peter (2016b): Envisioning the digital sustainability panopticon: a thought experiment of how big data may help advancing sustainability in the digital age. *Sustainability Science*. 11(5), 845–854 https://doi.org/https://doi.org/10.1007/s11625-016-0381-5

Seele, Peter (2018 im Erscheinen): Ab wann kann man (legitim) vom Digitalen Zeitalter sprechen? Eine Untersuchung aufbauend auf Hans Blumenbergs Konzept der Epochenschwelle samt einer historiographischen Typologie von Epochenwenden. In: Sölch, Dennis (Hrsg.). Philosophische Sprache zwischen Tradition und Innovation. Frankfurt am Main: Peter Lang.

Sennett, Richard (2003 [1976]): *The Fall of Public Man*. London: Penguin.

Smith, Martha J; Clarke, Ronald V; Pease, Ken (2002): Anticipatory benefits in crime prevention. *Crime Prevention Studies* 13 S. 71–88.

Wasserman, Elizabeth (2005): Rebels Without a Cause. *The Atlantic*, Technology, April.

Wewer, Göttrik (2013): Die Verschmelzung von privater und öffentlicher Sphäre im Internet. In: Ackermann, Ulrike (Hg.): *Im Sog des Internets: Öffentlichkeit und Privatheit im digitalen Zeitalter*. Frankfurt: Humanities, S. 53–70.

Woolf, Virginia (1977 [1929]): *A Room of One's Own*. London: Grafton.

Zapf, Lucas (2015): Martin Luther, Wealth and Labor: The Market Economy's Links to Prosperity Gospel. In: Heuser, Andreas (Hg.): *Pastures of Plenty: Tracing Religio-Scapes of Prosperity Gospel in Africa and beyond*. Frankfurt a. M.: Peter Lang, S. 279–292.

Zittrain, Dave (2011): Did Steve Jobs Favor or Oppose Internet Freedom? *Scientific American*. Abrufbar unter: http://www.scientificamerican.com/article/freedom-fighter/ *(letzter Zugriff: 07.04.2016)*.

Žižek, Slavoj (2006): Nobody has to be vile. *London Review of Books* 28 (7), S. 10–14.

Part II

Symptoms of the Structural Change of the Private

3

Showcasing Digital Omniscience in Everyday Life

3.1 Driving a Taxi

Digitalisation is changing the reality of life. It stands as a major social change above many everyday activities—and in some cases it is particularly noticeable. Like driving a taxi, for example, which has fundamentally changed with the emergence of smartphones.

3.1.1 "Hello Taxi!"

At the familiar signal—arm out and "Taxi!"—a Mercedes brakes. The car is easily recognizable, differently colored or patterned depending on the region, usually with a sign on the roof: TAXI. Transportation for a fee, performed by a professional, a historically proven business since the advent of hired sedan chairs in ancient times.

You get in. The interior: slightly worn leather upholstery, in the driver's area a photo or something else personal that identifies an individual workplace. After naming the destination, a brief nod from the driver. The city is unknown to the passenger and how far it may be to his destination. On a small display in the rearview mirror or on the center console, the taximeter relentlessly increases, accompanied by the skeptical looks of the passenger. With the maze of small and larger roads, there is nothing left to do but trust the driver to choose the shortest route. This is another reason why the choice fell on an

---

The translation of this chapter was done with the help of artificial intelligence (machine translation by the service DeepL.com). A subsequent human revision was done primarily in terms of content.

officially marked taxi. It is clear that the driver knows his trade, chooses the most direct route, has a clean and well-maintained vehicle. Trust in the functioning of this market (cf. Zapf and Seele 2015; Seele and Zapf 2017).

So the passenger has to have faith. He is at the mercy of an unknown person at the wheel in the small space of the vehicle. Whether the shortest or best route is chosen, whether the driver shows consideration for pedestrians, cyclists and other road users, the passenger has to accept all of this when he beckons the car. That's a risk you have to take, I suppose. If in doubt, the taxi insurance will pay if the passenger hits his head during a risky overtaking manoeuvre. More timid natured passengers will also worry about the driver delivering them to a dark corner, handing them over to waiting accomplices. After all, where the vehicle is and where it is headed is only known as far as you can see out the window. It might even be of use if the passenger might further think to himself, that not everyone knows where the taxi is currently travelling and how fast: 'As fast as possible to the airport' only works, after all, if the driver is not controlled by his own working equipment. And if the driver knows where there are speed cameras in the city.

The short conversation with the driver lasts as long as the drive. It revolves around the shops, the city, the car model. Every now and then the radio crackles, the control centre is asking for a car. The driver then turns down the radio, an answer is not necessary at the moment. He turns his attention to the next fare once the current one has been completed. In the end, the ride turns out to be unspectacular: the guest is brought to his destination swiftly and professionally, without detours, accidents or kidnappings. Payment is made in cash, the satisfied guest rounds up. The receipt is not necessary, the ride ends, the door closes. Driver and guest go their separate ways.

3.1.2 TaxiApp

The taxi app opens on the screen of the smartphone. The display shows a map of the surroundings; a few small taxi symbols are entered on the streets in the area. At the touch of a button, a car is ordered to the automatically determined location. Without researching the local taxi center, without thinking about the pickup location. With the convenience of the touch of a finger, the user chooses between different vehicles and drivers rated by a community system. Finding a taxi is no longer a matter of looking down the street in a trained manner and looking for the taxi sign on the roof of the car, but the rating scale of 1–5 stars on the smartphone display.

A click on a well rated vehicle ('clean', 'driver reliable'). Shortly thereafter, the message appears on the screen of the user: The taxi is coming. The unclear conditions in the taxi selection, should they once have given pause for thought, are eliminated. Transparency all around: No queasy feeling whether the driver is kidnapping passengers—with a 5-star rating, probably not. He even shows consideration for cyclists, which would otherwise warrant a one star deduction. The taxi app wants to combine the best of two worlds: Digital, the convenience of ordering, analog, the familiar safety of the transport medium.

But the app is only a symptom of other changes that have taken place in the taxi. A navigation device is clamped to the windscreen of the ordered car. On the small display, the passenger can (at least theoretically) understand where the car is currently driving. One could probably ask the driver to enter the destination into the device. Practically, this will be dispensed with—the driver knows the best route. But the presence of the GPS device still inspires confidence. The dot on the display moves, creating clarity and confidence about position and route. In U.S. taxis, displays are embedded on the back of the driver and passenger seats, at eye level of the seated passengers. The location of the taxi and advertising spots are displayed alternately, while at the same time the route is checked and the latest in skin care is communicated.

The visible box with the display is not the only GPS module on board. Somewhere inside the car there is another transmitter that transmits the driver's position, his speed and his breaks to the control centre. There, they know where and how the fleet is currently moving. The number of confidants privy to the position and the route of the vehicle is rapidly increasing. In addition to the driver (hopefully), the navigation device, the taxi dispatch center, even the taxi app provider now know where the passenger and driver are. The location of the hire and the route taken is communicated to the app. The fear of being hauled away undetected gives way to the uncomfortable feeling of being controlled. Statistics run along with all the gods behind the cloud: How often is the navigation system used, how long does the driver need for his routes, how often is a taxi ordered for which distances.

At the end of this transparent taxi ride, the passenger can give his rating. Everything is fine, four stars. One wants to leave room for improvement. Maybe, during his next visit to the city he will choose the same driver. After all, the data has been exchanged.

3.1.3 Uber

We are entering the next phase of digitalization: an online brokerage service for passenger transportation. The Uber app reports an impressive number of available cars in the vicinity—private cars with private drivers, for those with professional drivers, large cars, off-road vehicles, the offer is differentiated. With a flick of the finger, the app puts you in touch with a well-rated driver who is currently available. If it fits—route, pick-up time, ratings—the car is right there. The destination has been entered into the app beforehand, payment is arranged by a deposited credit card and is debited at the end of the month, including the 20% commission for the platform provider. When the guest gets in, the basic conditions have already been clarified. Including information on whether the driver tends to accelerate and brake jerkily or drives more smoothly and evenly. The taxi ride of the twenty-first century is beginning, with global success: according to the Wall Street Journal, the company is estimated to be worth US$50 billion in 2015. Founded in 2009.

And all cheaper than a classic taxi. The price is flexible and is set by the platform: Live, depending on supply and demand. Many free drivers and few passengers: low fare. Few free drivers, high demand: rising prices. The algorithms calculating in the background on the smartphone collect data continuously—where however many requests are answered, where it goes, on which route, who drives, who rides, at what speed, where stops and turns are made. The data ends up on the Uber servers, where it is analysed (cf. NDR 2015). Another popular option is to let your networked friends know directly from the app where and how you are currently on the road. Then everyone knows.

A blog entry by Uber from 2012 shows the potential of this collected data even back then. Under the title 'Rides of Glory', it was calculated in which US cities most one-night stands take place. This was done by filtering rides for two people in the late night and early morning hours, as well as single rides in a small radius around the drop-off point a few hours later. According to this research, Boston may consider itself the capital of this type of pleasure (see Voytek 2012). The blog entry could be found on the homepage for a few years without much fuss, until it was removed by Uber due to the negative reactions of data protectionists.

'Rides of Glory'—and its deletion—provides a brief moment's insight into the mindset of the data-collecting taxi service. Data collection and processing, even in relation to customers' most personal activities, comes so naturally to the company that the results of the investigation are published on its own blog. To the point where they generated too much negative attention. Then the post was deleted, the study described by an Uber representative as 'fun with data'. And the data collection quietly continued from then on.

The metamorphosis from private car owner to Uber partner is completed with a simple online registration. The platform entices with the opportunity to earn a little extra on your own. The earning potential is good—partners in the USA earn up to US$1200 per week if they mainly do Uber rides. With no boss and no set hours. However, dependent on the Uber platform, which sets the share drivers get from the fare—and occasionally cuts it for its own benefit. Drivers who protest against this risk being deactivated as Uber partners.

The journey itself is unspectacular, swift and professional. The logging in the background imperceptible. The destination is reached, but the journey is not quite finished when the driver leaves the vehicle: The assessments are still due. Driver and guest are dependent on a positive assessment of the other, the community wants to know how the mutual driving experience was. Drivers and passengers agree: *every time again*—safe, friendly, clean, on time. Uber will also be happy to hear that.

3.2 Overnight Stay

Little is as private as sleep. Secluded, you spend the night completely with yourself.

Sleep may be private, but its circumstances are not necessarily so: the spectrum from the private bedroom to the park bench illustrates this. Circumstances of sleep can aim precisely to be experienced by strangers and not to be private. A performance by Julius von Bismarck, shown at *Art Unlimited in* Basel in 2015, comes to mind. The artist positioned himself, along with a desk and a mattress, on a spinning disc in the middle of the exhibition. On this rotating living space the artist stayed, worked and—slept.

The circumstances of sleep reflect social developments. Who sleeps where and with whom? Projects such as *Kommune 1* in Berlin in the late 1960s took up this connection between the private, the public and the prevailing social ideas: The political residential community, sleeping together on a mattress dormitory became a public event, the commune celebrated by the initiators as a "destruction of the private sphere" (Matussek and Oehmke 2007, p. 140).

Along our theme of the secret private, we follow the circumstances of sleep by means of a three-step process: from the bourgeois bedroom to the boarding house to airbnb accommodation.

3.2.1 The Middle-Class Bedroom

Draw the curtains, lower the blinds. With this act, performed a thousand times, the secret private comes into the bedroom. Facing the quiet side of the house, a room of almost mysterious quality is created. Visitors are not necessarily led through the bedroom, if at all the door is briefly opened during the obligatory tour, a glance from the outside must suffice. What happens here behind closed curtains is nobody's business.

Today, the protected bedroom is part of living culture in the West. However, such a room reserved for private sleep has only been common since the eighteenth century, and a bed of one's own only since the beginning of the twentieth century (cf. Pollak 2013, p. 20). Before that, people slept together in large beds, naked except for the obligatory nightcap (cf. Mohrmann 2012, p. 24). This development is also an expression of moral customs: Curtains and the closed door, the partitioned bedroom and clothed sleep are linked to Christian sexual morality. Sexuality belongs only in the most private space, outsiders should not hear about it (cf. W 2015).

The reduction to a maximum of two adults in one bed is related not only to moral considerations but also to the increasing availability of living space. In this respect, private sleep is a consequence of increasing prosperity, social departure from the collective and a more individual way of life (cf. Mohrmann 2012, p. 26).

Thus, the bedroom presents itself as a bundle of social ideas, shows the social and economic status, moral ideas. Last but not least, it also reflects the marital status and more precisely: the arrangement of a relationship. Separate beds, separate bedrooms, a large bed with two small blankets, a large bed with one large blanket or the single bed. In all cases, the bedroom is a private space whose inner workings are the secret of its regular users. The classic bourgeois bedroom spatially accommodates the individual personality and its need for privacy—be it shared with a partner.

3.2.2 Overnight Stay in the Boarding House

An exhausted traveler sees a boarding house at night on the side of the highway. The sign in front of the entrance lights up green. Free rooms! The traveler pulls into the parking lot and enters the building. From a room behind the reception counter a somewhat sleepy clerk emerges. The price is mentioned, it is ok. A brief personal contact, filling out the registration form, then the room key is handed over with its heavy brass fob. Breakfast from seven to half past eight.

You do not want to know exactly, at most you suspect how many hundreds of people have already slept in this bed. The sheets may still be fresh, gleaming white and starched. The bed in the boarding house is a strange bed. A borrowed place to spend the night, a realm between the very private and the public-economic. Until the seventeenth century, people slept in boarding houses with four or five strangers in three beds per room (cf. Mohrmann 2012, p. 24). Today, of course, the guest is to himself—a private room for one night before the next guest closes the heavy curtains and turns the room into his short-term, secret private.

The room is furnished as always, the bed a bit too soft, the TV a bit too small. At breakfast the bread is fresh and the cheese without much taste. Payment of the room is preferably in cash, credit cards cost too many fees. From the night remains a casually stowed bill, a registration form, of which remains puzzling whether it is forwarded to the competent authority, little else. The memory of the overnight stay will soon be gone, just as the visit will not occupy the experienced hostel warden for long.

3.2.3 Airbnb

Airbnb mediates private overnight stays, opens private apartments to the internet. Overnight stay against payment in a hitherto private space. The host opens his apartment to a stranger, the only connection is the platform. Perhaps one has previously seen some photos that the system provides.

The distant handing over of the keys at the reception is not the prelude to the typical Airbnb overnight stay, but the appointment with the host. The private contact is mandatory—after all, you deliberately don't want to sleep in an anonymous hotel: "Welcome home" is the company's slogan, with "unique accommodations from local hosts". The guest stays in a private atmosphere with local colour and that certain personal touch, in student shared flats where a room is just available, in large city apartments or houses of people who are interested in travellers, who like to show off their apartment and earn a little extra money. The welcome is friendly and personal, the short tour of the apartment afterwards is exciting: who doesn't like to look into the kitchen and living room of strangers? If the *local host is* ambitious, has special rooms, is carefully prepared and gets good reviews, he can charge as much as a medium-priced hotel for the overnight stay.

Anyone who wants to rent or let a private room online is quickly faced with the problem of trust, or more precisely, a lack thereof. Because the disembodied and voiceless platform through which the sleeping place is arranged initially creates a queasy feeling on both sides. What if the guest turns out to be a kleptomaniac ruffian, the hostess less interested in my night's rest and more in my luggage? These are the obstacles that Airbnb counters in ingenious ways: Customers—*host* and *guest*—are registered, links to other social media activity are expected, as is the submission of an ID scan. And, central to building trust: reviews by other users about previous overnight stays. Additionally, the host is insured through the platform against damage caused by guests. For this, and for the fact that he can use the platform, about 10% of the accommodation revenue goes to Airbnb. The contact between guest and host runs entirely through the operator's communication system, as does the subsequent evaluation of both parties. Guest and host have, as long as both use the platform regularly, an interest in a good rating. A part of the rating is public, another part is only visible to the operators of the platform. Airbnb closely follows the initiation, implementation and completion of the accommodation: Room preferences of users, passport data, Facebook account and pricing options are known, as well as the type and extent of communications. How the room is left in the morning is stored via the hidden ratings. If we recall the definitions of privacy made at the beginning (see Sect. 2.1), one of the aspects was the *absence of valuations* that constitute privacy. For those who use Airbnb, this definition no longer applies.

Payment is settled directly via the guest's credit card and the platform. Cash would disturb the atmosphere, which lies somewhere between private room brokerage, internet publicity and entrepreneurship.

3.3 Celebrating and Eating

Eating and celebrating together is accompanied by a variety of rituals, social and unwritten rules of interaction, etiquette and exchange. As social and cultural practices, they are subject to fashions and developments: From the Viennese court ball to the flat rate party in the village disco, from the Sunday roast in the family circle to the online streaming of the singles' dinner, this becomes clear. The familiar is combined with the new, discarded again, expanded. Celebrating and eating together also exists in varying degrees of privacy; consider the range between a romantic dinner for two and an *open house*.

In the following, we describe contemporary forms of eating and celebrating in which the structural change of the private in the digital age becomes visible: The student *get-together* where everyone brings something to eat, the running dinner as a semi-public meal, and most recently forms of digital sharing of food.

3.3.1 'Everyone Brings Something'

At family gatherings, sometimes we had these little apple pies for dessert. Then it was a good family celebration. An apple ring, fried in pancake batter. Recooking it, the taste level never compares. So a call to grandmother's house is necessary, after careful banter the secret is extracted: two tablespoons of vinegar in the batter, who knew. The milk acidifies, the dough becomes looser. Good, I'll try it out right away, a few fellow students are coming over for dinner tonight.

Everyone is informed, a low-stakes event without a lot of planning. You know each other, everyone brings something. If necessary, also a friend, but not too many strangers. The selection of the evening: almost all students, almost all from the same course. A closed event in one's own four walls, supported by a personal, familiar circle of people. Communicative, cooperative, with a tendency to eat too much noodle salad.

Is this an expression of a 'generation of stinginess' that only provides its friends with an apartment, but otherwise avoids any cost consequences of social interaction? Probably not, the goal is different: it's about being together, not about cementing the status of the host with formal invitations and expensive food (at least that's what research says, cf. Büchel 2015). Not to mention the complications that arise by cooking for a circle of friends representing the entire dietary spectrum from exclusively meat to vegan.

When everyone has eaten, drunk and talked enough, you go home, party's over, until the next time in the next shared kitchen. At the end, everyone can take some of the leftovers—why didn't anyone try the almond milk soy-based cream cheese? The apple pies are all gone for that matter.

3.3.2 Running Dinner

Running Dinner is an event where you cook food and meet new people. In teams of two people, a total of three courses are prepared, each of which is eaten in a different apartment. So you change locations three times and eat with the cooking team of the current course. The event is planned by an agency: Who, where, what. The focus is on getting to know previously unknown people. The private home is opened to a limited but unknown public—those who attend the event. A dinner in private, but with strangers. The name 'Running Dinner' is now a registered trademark, events of the same type also continue under a different name ('RuDi'). The composition of the cooking teams and the entire dinner party is largely random, determined mainly by who has signed up for the event. The common meal as a platform for exchange and getting to know each other. The party is made up of six people at a time, each unknown to the other, some of whom open up the private space of their homes to this limited public. Not a closed event like the bring-your-own-party, but openly accessible, but still within your own four walls. Let's get started.

In the evening, the cooking event is planned, for the *running dinner* you will make the appetizer. Now, what to prepare? Normally you like it simple and fast, *Insalata Caprese* or similar, little effort. But that's not possible today. Because of the unknown guests, it has taste as good as possible for everyone. So, you turn to the infallible taste of the masses: today chefkoch.de is chef, so purely in the appetizer section. Among the 26,146 recipes entered, the undisputed leader—4,855,436 views—is pumpkin soup with ginger and coconut milk. Nearly five million amateur cooks and a rating of 4.64 stars can't lie. The recipe has already been printed 423,903 times, the statistics under the recipe proudly report. How many trees have probably been used for the coconut soup? Therefore, you only download it to your tablet. How practical, at the touch of a button, a shopping list can be created, which is delivered to your home by an associated company if required. For around 20 € in the case of the soup. If desired, optimized for organic ingredients (16 €). Only the best for the *running dinner*.

3.3.3 Food and Party

The most digital way to cook? The Vorwerk Thermomix! One salesman's evening and eleven hundred euros later, the miracle appliance is in the apartment, the food processor of all food processors. With a touchscreen, the device opens and closes automatically and has an interface to which a storage medium with recipes is clipped. The ingredients are put into the machine in sequence according to the instructions on the Thermomix screen, chopped, stirred, heated: "Here, in a word, the algorithm cooks" (Dworschak 2015, p. 111). The potential seems huge: the networked kitchen of the future will order and cook on its own, even the ingredients can be prepared in the medium term by a robo-cook, a device that essentially consists of two arms, including washing up—available from 2017 (cf. Meusers 2015). The by product of this networking is detailed access to the dietary habits of kitchen and appliance users. What, how much, how often, shopping lists can then be created accordingly, product recommendations can be personalized—or warnings issued: *Mousse au Chocolat? That doesn't have to be there anymore! Please take a yoghurt from the fridge.*

Digitally, it moves seamlessly from the kitchen to the dining room. *ShareTheMeal* is an app of the *UN World Food Programme* WFP. With one click, users donate 40 cents in the app, which is exactly the amount WFP says is needed to feed a child for a day. To donate, personal information is needed and use of the app is logged, according to the providers' homepage. Where people donate from and how much time is spent on the app is recorded to improve the user experience. So you dine, perhaps in Munich or Milan, seemingly alone but virtually connected to a child 5000 km away. Abstract and at the same time close to life. You eat alone and yet you are not.

The abstract sharing of food seems to be a trend in general. Less charitable than the UN, we encounter it in *food porn*. Just been served a particularly good-looking meal? Don't be shy: pull out your smartphone, take a photo of the deliciously laden plate, write a snappy sentence about it, add the location of the photo and upload it to the social network. This is liked and shared, commented on, 'Great!', '#nomnom!', or sometimes even critically 'Does it always have to have meat in it?' All while eating alone, privately in front of the plate, and yet connected to a global audience that virtually witnesses the consumption of the pasta that has just been photographed. Sean Garrett, a US-based PR man with a focus on the digital, even believes that the image of the meal was just the beginning: "Early Twitter: 'What I had for lunch.' Early Meerkat: 'Watch me eat my lunch'" (Garrett 2015). So he streams his food live as video to the internet. Whether with the photo-based tweet or the smartphone-based video stream, it is eaten, documented, and shared. Yesterday's live food tweet, today's live food feed. The private meal as a public event, that's what works about food. And, as we'll see in a moment, it works even better when celebrating.

First, let's take a step back, away from the digital. Let's start with the disco. The visit is initially anonymous. A public place, the encounter with other people rather unplanned and open-ended. The person celebrating is distinguished by their physical presence, visible but unannounced. So far, so analogous. The visit is quickly forgotten if things don't go well, in case of doubt the last liquor helps to blur unpleasant scraps. But wait, not so fast. There were photographers at the party, professional ones. Sent by external agencies, they were called night agents, scene1 or guestlist030. They were taking party photos (being a party photographer was suddenly an occupation). The party photos prevented the evening from ending when they stepped out of the venue. The photos were uploaded to the agency's website that same evening, online for everyone to see. The photographer had distributed the flyer with the web address on it on the dance floor. And so, the next morning, the night could be reviewed on the screen at home. Endless amounts of generic photos, young people with red heads and glazed looks, arm in arm with friends laughing into the camera. First you looked for pictures of yourself and then of your companions. Ah, him too, but I was already gone, why is she in the photo with him?

Party photographers were only needed in the short interim period when digital photography had established itself, but the devices were not yet so cheap and small that everyone could have one with them at all times. For a short time, money was earned this way: With the party pictures on the internet, attention could be drawn to the homepage. Other products were then advertised or offered on the homepage. In today's digitally networked world, this has not changed and even works better across the board: nowadays, the photos are taken by the visitors themselves and uploaded to the various private profiles of social networks, where they can then be viewed, commented on and shared instantly by other visitors. The principle remains the same: users generate attention with private content, which is used by platforms for business purposes. The private *in the public* becomes an economic object. And the party photo was just the beginning.

We continue with the PartyGuerilla, from the disco directly into the living room. Now the private *in the private* becomes an economic object. With a piercing idea: sponsor WG parties and use the party as an advertising platform for drinks and new products, a lived personal recommendation, emotional brand loyalty in a private environment. The agency behind PartyGuerilla uses the potential of private networks to bring brands closer to consumers. The name part *Guerilla* makes the underlying strategy transparent: Irregular warfare, which is characterized by high mobility of the fighters, elaborate concealment and obscuring tactics and consists of numerous small, powerful actions. The battle we are dealing with is the battle for the potential customer's attention. That attention will be drawn to the product by any means necessary—*party!* And it works like this: an interested student applies to the company. If he and his party concept are selected, Partyguerilla provides the *product support* (i.e.: supplies the drinks), and the applicant in return gets access to the circle of friends. The party runs, there are branded drinks, which otherwise would not be financially affordable. *Product placement and emotional brand loyalty in a private environment*—with success control, as it says in the general terms and conditions: number of guests, atmosphere, duration of the party, the key data of the event are analysed. The *Student Brand Manager* provides *real-time product feedback* directly from the scene of the action, from the *Real Life Market Research of* the party. This is what the agency offers:

> By accepting the free merchandise for your party you agree that a Student Brand Manager from your city will be invited to your party and will be on site for documentation. Our Student Brand Manager will check if the information you provide about the party is true and take pictures of your party [...] (Partyguerilla 2013).

The party aspect of the concept is emphasized on the homepage, accordingly everything is colorful. *Partyguerilla* advertises the party concept on an independent homepage. The business case behind it, the contact point for the advertising companies, is presented on another domain: btl-creative.com, business-like grey and informative. The two worlds—the private and the economic—are initially kept apart in the virtual—and only brought together in the product, the sponsored flat-sharing party.

Partyguerilla reports that within 2 years more than 200,000 students have enjoyed their parties and thus come into contact with the advertised products. Out of the student party with cheap beer and Aldi salt sticks, into the advertising-financed brand party. Including live target group research and brand loyalty. You remember this particular beer with particular fondness, you remember it from that *legendary* flat-sharing party where you met that likeable fellow student, what was his name, that was fun. Or, even more better from a brand loyalty point of view: Do you remember, darling, the party where we drank so much Absolut Vodka and then I finally dared to. That's probably how brand loyalists imagine it, or something like that. And that's how it seems to work, because for the love of student life, the various companies would hardly bottle 200,000 students.

With another product of the same agency—*TrendGuerilla*—the party concept is extended to smaller events: Beauty night, Games night, Cooking night, Champions League night, any form of private celebration can be catered for with products. "Unique—Unforgettable—Authentic—Among Friends—Emotional—Familiar" (btl creative 2015). The student target audience is particularly suited to this, as there tends to be sufficient time and a whole range of different ritualised forms of social encounter available. And if the trend and party guerrilla has done its job well, brand loyalty in the milieu of the (hopefully later) high earners is assured.

Partyguerilla mixes personal recommendations and the positive experience with a brand into an advertising concept. The targeted entry of the economic into the private sphere. Planned as a commando operation by an advertising agency. Capitalism and war as two sides of the same coin, Sara Wagenknecht had the right nose (cf. Wagenknecht 2009).

3.4 Sharing

Sharing—in the sense of sharing—promotes the joint use of something that was not shared before: 'Yours' and 'Mine' become 'Ours'.[1] The product is transferred from individual to collective use. After the act, what is shared is accessible to more people than before; sharing involves social contact, a connection and a certain relationship to the community.

Sharing presupposes an awareness of 'mine' and 'yours'. It happens voluntarily, otherwise it would be stealing or taking away. It is not a trade—for there is no material consideration. It is not a gift or a present, for a part remains with the original owner. It is something bigger than trade in objects, for it is not limited to the material. Accordingly, a distinction can be made between the sharing of tangible things (e.g. a loaf of bread) and intangible things (e.g. a piece of information). In the first case, the nature or availability of what is shared is changed; in the second case, it is not.

Behind sharing there is usually a conviction, perhaps even a whole worldview and idea of being, or, deep down, at least an ulterior motive. These ideas behind sharing are, as we will see in the following chapter, subject to change—induced, for example, by the digital push into the analogue, or by the economic push into the social sphere.

3.4.1 St Martin: Sharing out of Religious Conviction

St. Martin—the name symbolizes sharing like no other. And with an ulterior Christian motive. Martin of Tours, officer of the Roman army in the fourth century, later bishop, and then saint of the Catholic Church. We owe the

---

[1] Another kind of sharing focuses on separation: *dividing* into 'your half and my half' in the sense of the English *divide*. This sharing is a compromise: according to it, the sharer owns less of the shared good, but he has achieved a purpose. This sharing focuses more on the shared object than the use, thus less on the act of sharing.

tradition of his now proverbial desire to share to a half-naked beggar: Martin, on horseback, rides out of the city gate. There he sees a man without clothes asking passers-by for help. No one stops, the freezing man is ignored. Martin feels sorry for the beggar, it is cold and the figure looks pitiful. But Martin carries only his weapons and uniform. So, under the mocking looks of the passers-by, the soldier cuts the cloak of his uniform with his sword and hands one part to the beggar. The other part he puts on himself again. And rides on. Apart from the reduced warmth caused by the division, this is not entirely unproblematic for Martin: the soldiers' uniforms were the property of the emperor, and destroying them was punishable by imprisonment. Whether Martin merely froze for his sharing or was also imprisoned is not entirely clear. In any case, the deed does not go unnoticed: on the night of sharing, Jesus appears to Martin in a dream. And Jesus wears half of the cloak (Fig. 3.1).

Countless statues and portraits bear witness to the meaningful scene of sharing that is said to have taken place near Amiens in the north of France and which, according to Martin's own account, so lastingly affirmed his faith and its expansion (cf. Wachwitz 1962, p. 780). Martin, at the moment of sharing not yet a 'St', cuts his coat in two against the background of his charity, his love of his neighbour, which obliges him as a Christian to act. The deed, his dream shows, finds favour in the highest place.

The story is a classic, designed to teach children the joy of sharing. The generous act, performed by the proud soldier, is done out of compassion and

Fig. 3.1 St. Martin of Tours at Basel Cathedral. (Photo: Meskens 2010)

the conviction to alleviate the suffering of a fellow human being. St. Martin may be cold now, but the beggar is a little warmer. Martin shares out of selfless and religious motivation. Made possible by an open, attentive eye for those in need of sharing. The looks or comments of others are irrelevant. Martin shares out of private conviction, despite the surrounding scoffers, contrary to his role as a soldier in the Roman army.

3.4.2 Collaborative Consumption: Sharing for a Better World

St. Martin was motivated to share by his Christian faith. If we jump into the networked modern era, we find similarly motivated sharing, but linked to social concern instead of religious enrichment: It is meant to foster community, to critique blind consumerism, to reinforce an environmentally and socially conscious attitude. Sharing with the goal of social change.

The politically motivated 68er 'shared apartment' comes to mind: sharing living space with friends or strangers as a society-changing act. Together with the outdated moral concepts, the closed nature of the private sphere is broken up. Of course, the bedroom is also shared, and cases of toilet doors being unhinged come to light. Sharing as a political act, the private as the political, living as a demonstration of the will to change.

What remains today of the politically motivated commune of '68 is mostly the purpose-built shared apartment: Living together because it would be too expensive alone, not politically but economically motivated. The WG residents have their own separate bedrooms. The bathroom has a door again and shared living has become mainstream. Sharing in this sense means sharing the product and its costs with others—rather for pragmatic reasons and without lapsing into greenwashing (cf. Seele and Gatti 2017).

Or the shared car. The shared vehicle, well-organized with insurance and shared responsibility for refueling and washing, based on the conviction that the car sits around a lot, is expensive, and the running costs are high. So they use it together, save money and do something good at the same time. In this way, disadvantaged population groups benefit from having a car, fewer of them are needed overall, and traffic decreases. Sharing based on social considerations and environmental awareness.[2]

---

[2] As we will see, car sharing has now entered the mainstream of the car industry—with a slightly different concern. We turn in the next section to the *collaborative* or *sharing economy*, which focuses on the business case for sharing.

Based on the desire for social change, private property is put on the back burner. *Collaborative consumption* is the associated keyword of this discourse, which has been steadily gaining importance since the 1980s, judging by its prominence and the number of initiatives. In this concept, the demand side is at the centre of attention, together with the notion of 'good' shared consumption. Cooperative models—for example, in building houses or purchasing agricultural machinery—have been around for a long time. However, the extension to other consumer products and the link to an awareness of social change came into the consciousness of a wider public with the idea of the shared car. Products are now shared and consumed, thereby inducing and consolidating a change in social thinking. The existing free-market system is to be maintained, only certain developments perceived as harmful are to be opposed. In their book *Collaborative Consumption,* Rachel Botsman and Roo Rogers provide a historical classification and substantive orientation for this change:

> A] sustainable system built to serve basic human needs—in particular, the needs for community, individual identity, recognition, and meaningful activity—rooted in age-old market principles and collaborative behaviors. Indeed, it will be referred to as a revolution, so to speak, when society, faced with grave challenges, started to make a seismic shift from an unfettered zeal for individual getting and spending toward a rediscovery of collective good (Botsman and Rogers 2011, p. 209 f.).

In this concluding sentence of the book, the reader gets a glimpse of the greatness of the idea that is being negotiated. For the economy, nothing less than 'the collective good' is rediscovered. It is about changing consumption and rethinking the economy, about personal restraint. Progress for the environment and society: sustainability increases, resources and not least money are saved. At the same time, collaborative consumption serves "fundamental human needs" such as community, identity and meaning. The economy is no longer guided by individual well-being, but by collective well-being. Sharing is fundamentally revolutionizing society. And, according to the authors, it comes without ideology or dogma (cf. Botsman and Rogers 2011, p. 18). Connected to this revolution is an economy that creates community, meaning and identity through sharing. This clearly goes beyond the basic economic functions of exchange and allocation. With this charge, the economy becomes the central organizational variable of social life and society, and economic culture becomes the culture that determines society. This statement relativizes the claim that *collaborative consumption* is free of ideological approaches. On closer examination, the opposite seems to be the case. Private consumption becomes a bogeyman, an environmentally destructive, selfish act. Private use is subject to the latent accusation of hyperconsumption: To get into one's own car and drive as long as one likes, to use the washing machine as often and as long as one likes, to drill a hole with one's own machine once a year. One looks in vain for shades of grey in this form of sharing. Instead, the private nature of consumption is suspended.

This reassessment of economy and sharing would hardly be possible without digitalization, without the practically cost-neutral networking and the associated exchange of information. Neither would the necessary critical mass of participants come together, nor would sharing be sufficiently simple and convenient. Within the sharing community, the use of the shared products must be transparent. Shared consumption therefore relies on the entire arsenal of electronic sensors: Pictures of the shared accommodation, live location of the shared car, location of the nearest free shared drill. As this information becomes public, the private is reduced in the use of shared products. Who is staying where, for how long? Where, how often and by whom are the car and drill used?

In the analogue world, the acquisition and use of products was something individual, private. With the reduction of individual ownership and collective consumption, private use becomes a public act. An active choice against the private. And that is just the beginning.

3.4.3 Sharing Economy: Sharing as a Business Case

What began as *collaborative consumption* on a small scale at the consumer level is being rolled out on a large scale, professionalized, optimized: Sharing for a better world becomes a business case. Sharing becomes an entire sharing economy, the very big business case.

The development of Car-Sharing is an example of how this business caseification behaves, how the ulterior motives of sharing are transformed: An alternative project driven by social conviction has now become a full-fledged revenue stream in the automotive industry. Large car companies offer their vehicles for flexible rental in cities. The car is found, started and billed using an app. An alternative to local transport. In addition, the cars are visible in the cityscape, potential customers can test drive without strings attached. Drivers become accustomed to the vehicles and innovations such as electric cars being placed close to customers in everyday life. Travel time, travel distance and pick-up and drop-off locations are recorded and evaluated (cf. civity 2014). In the urban environment, where car companies compete with bicycles and public transport, this results in an additional sales market for the automotive industry (cf. Schlesiger 2014).

Sharing as a business case relies on digital infrastructures for distribution and collaboration opportunities. The decentralized organization relies on easy access to sensors—such as location and the possibility of identification. Sharing becomes professionalized and can be optimized thanks to the generated data including its analysis. With an increasing degree of digitalization, the range of sharing is also increasing. Digital platforms are establishing themselves, some of which now have the revenues of large companies. The auditing firm PWC calculated global revenues of US$15 billion for the sharing economy in 2014. By 2025, this figure is expected to reach 335 billion (cf. Hawksworth and Vaughan 2014).

The consulting firm *Crowd Companies*, which supports companies from the sharing economy, lists 12 major areas of activity from this field: from learning to money, health, food, transport and logistics, the concept covers a broad field (cf. Owyang 2014). Many of the companies involved are now globally known brands: Bitcoin, Kickstarter, Etsy, eBay, Craigslist, airbnb or uber. Users are not merely customers who buy a product from the company. Rather, at the heart of this industry, people are seen as equals to the company, acting as "Makers, Co-Creators, Crowdfunders, Peers, Customers" (ibid.). Users of the digitally-enabled sharing economy have a more active role than the old, passive and analog consumer. Digital consumers participate in business models: They participate in mining a currency, act as investors with small amounts of money, sell handmade items or auction used consumer products that are no longer needed. They offer surplus living space to guests or drive other users through the city in their own cars. The digital consumer lives a dual role, exchange in the economy no longer works in only one direction; from company to consumer. The consumer himself becomes an entrepreneur, while at the same time consuming a platform's offerings through his connection to it: The consumer is both driver and passenger, host and guest at the same time.

The digital business case changes the rules of the game of sharing: In the logic of the sharing economy, this is now linked to financial compensation.[3] The car, the overnight accommodation, for all shared objects a financial compensation is agreed upon and additionally a share is paid to the intermediary. What a development: the analogue Martin shares his coat out of inner,

---

[3] In macroeconomic terms, the sharing economy is described accordingly as an area of innovation within the rental sector: Cars, accommodation, media (music, films), tools and equipment (see e.g. Hawksworth and Vaughan 2014).

religious conviction. St. Martin 2.0 receives a request for warm clothing on his smartphone. He shares his decision while still on horseback via Twitter. The part of the coat is given to the beggar in the vicinity who has the best ratings ("always thanks kindly—anytime again"). That's how sharing works today.

3.5 Tying up

When two people get to know each other, this initially only affects the two people who are involved. And yet the process clearly reveals circumstances that go beyond the two people directly involved: How and where do the two become a couple, which relationships are accepted, which sexual morals prevail, which role do family, religion and the state play. Friends and relatives, acquaintances, all witness the initiation, comment, give advice or silently form their opinion on the impending relationship.

To help chance along, there have always been institutions dedicated to initiating partnerships. From matchmakers to Sunday celebrities to online side jump portals, new forums for getting to know each other are emerging over time. Particularly visible from the outside, partnership or marriage agencies crystallize what the *state of the art* looks like. After all, the professional initiation of relationships needs to be advertised, the clientele needs to be convinced. In particular, the legal framework in which this business takes place is a sensitive indicator: matchmaking, as the violation is known, is no longer a punishable offence in Western societies (as long as it does not involve children). Nevertheless, the issue is not *business as usual*. For example, financial claims arising from a marriage brokerage are classified as an obligation in kind in Germany. Thus, they cannot be sued for, but must be paid voluntarily. Justification: Spreading the details of the marriage initiation in court is considered unreasonable by the legislator (cf. Reimer 2015). The initiation of a marriage is thus placed above the financial interests of the dating agency with reference to privacy.

Below we look at three institutions that serve to initiate a relationship: A traditional village festival, speed dating, and an online dating agency.

3.5.1 The Village Fete

The village festival, whether it is for the crowning of the new shooting king or the Dance into May, takes place on the market or festival square of the village. Mostly under the open sky and with tents it is an event in the public space. If you want to meet someone in this space, all you have to do is show up at the specified time, and everything else is based on appearance, habits, maybe even recognition from a previous event. And on the tacit agreement that at events like this, a basic openness to getting to know someone can be assumed—at least if you're travelling solo. Folk symbols like the position of the bow on the dirndl are meant to make this willingness efficiently visible: Tied on the left, on the right, in the middle or the little bow in the back. Depending on the regional tradition, the availability, secrecy or disinterest in the search for a partner is expressed in an elegant manner.

Whether the appearance meets with interest or even love is subject to chance, getting to know each other at the village festival is contingent: whether the partner for life is sitting in one of the tents is open. Maybe you catch a glimpse of him or her while walking through the festival grounds, maybe he or she is sitting at the same beer bench, maybe he or she is working behind the deposit return counter. Regardless, the number of potential partners is limited: By gender and age, the size of the locality and the festival grounds. By visiting the festival, the search for a partner is narrowed down locally. A selection without technical aid, purely through the presence on the festival grounds.

If there is a spark between the two, if they get closer to each other, the new togetherness is of course noticed. Attentively or by chance, the present entourage observes the scene. The village eye is present during the flirtation. At least as long as the two don't decide to leave. The short walk around the tent is private after all.

3.5.2 Speed Dating

We leave the traditional folk festival, because the urban dweller wants to escape the small circle. That's why he has swapped the narrowness of the village for the anonymity of the city. This presents him with a problem: the contingency in the choice of partner has multiplied, but how to exploit this huge potential? A gap in the market. And so, in the late 1990s, Rabbi Yaacov Deyo dreamed up speed dating. He was looking for a way to stop the declining birth rate within the Los Angeles Jewish community. And came up with the following concept: A fixed date in a café, an equal number of previously registered women and men. Every 10 minutes, each participant meets everyone of the opposite sex. At the end of the meeting, the participants tell the organizer which of their conversation partners they would like to meet again. If this is the case for both conversation partners, the contact details are released. If this is not the case, the names remain unknown (cf. Kennedy 2013).

In the semi-public space of a café or a disco, the contingent big city is reduced to a small group of highly motivated love seekers. Nevertheless, a hit of unpredictability remains, because everyone talks to everyone else at some point. The setting is clear but still anonymous: the identity remains secret at first. What is revealed in the conversation can vary with each interlocutor. If the right person is not there, it will work out on another evening. In the organizer's customer file, the contact details are noted, perhaps also how often the event has already been attended. Those who no longer come to the events leave no further traces.

In the process outlined, the seeker becomes the customer, the search for a partner becomes a business case—as far as this is known, since money has been earned with matchmaking. What is new is the degree of rationalization. Yaacov Deyo was concerned with using streamlined organization to identify marriage-minded partners and make the process of hooking up more efficient. Speed dating casts this in the form of an event: existing resolution and booking participation, then the systematic questioning and processing of interlocutors.

3.5.3 Parship

The Internet seems to eliminate several problems of analog dating. First, the selection is global, almost infinite. Secondly, the selection can be precisely filtered according to one's own ideas—appearance, location, preferences in films, music, education, religion, humour, there are no limits to differentiation. In order to cope with the amount of data, online partner agencies use an algorithm to decide which pairings seem promising. Based on the profile information, suggestions are made. The calculation of the 'match' is, of course, psychologically and mathematically sound, it exceeds the possibilities of human search, sorts through the crowd, makes a selection and gives confidence that the suggestions cover an optimum from the largest possible pool—and large it is: in Europe alone, more than 49 million people visit dating websites every month (cf. Bridle 2014). Another efficient advantage: the seeker no longer has to appear in person to advance the search. Unlike village fairs or speed dating, the profile is online and accessible 24 hours a day—there is no smarter way to find the right partner.

The partner search is not bound to a specific place or time and can be adapted on an ongoing basis. This increases the planning effort, data maintenance is essential for the image that is created by the profile of the potential partner. With constant optimization, the partner search reaches a further level

3 Showcasing Digital Omniscience in Everyday Life 75

of rationalization through digitalization, which is characterized by a latent evaluation of one's own and the desired profile in addition to intent and systematic procedure (speed dating). Fuelled by the conviction that *the perfect* partner can be found somewhere out there among the seemingly endless participants, the search for the consummate *match* becomes a game: individual participants use several profiles to see which one comes across best. Participants in a relationship leave the profile online to possibly get even closer to the perfect match.

The openness and creativity in dealing with one's own data as well as its publication stands in contrast to the seclusion of life in which this search takes place. In the personal environment, the participation in online dating remains clandestine. Unlike the conspicuous visit to the village fair or the date for speed dating, online dating goes unnoticed. One's personal circle no longer knows about it. But there are new confidants: the operator of the partner exchange. When which information is entered (spent the whole Saturday evening alone in front of the profile again?) and whether this information matches the user's other measurable behaviour: Where is the user surfing on from the partner page, how and how often is communication with potential partners taking place (it's not sparking properly yet) and what deviations are occurring? Everything is measured and logged, all in the service of the perfect match.

The site operator knows which partner fits, even before the two partners know it. Especially with dating agencies that specialize in non-traditional or extramarital relationships, this knowledge is quickly explosive. Such is the case of the infidelity portal *Ashley Madison*. Well aware of the latent guilty conscience of its clientele, the platform offered to erase all personal traces of the user from the site for a fee of US$19. Mind you, as a paid extra service. The promise to make infidelity invisible and banish it from the online public sphere went brilliantly, earning the company additional millions (cf. Bernstein 2015). But now the portal's customer file was recently hacked and made freely available for download on the Internet. And it turned out that the data had by no means been deleted. With the help of a simple online form, unsettled partners could now use their e-mail address to check whether they had been cuckolded. In the Ashley Madison case, the hackers who made the logging of the partner search and sex life public do not seem to have pursued financial interests, but—at least pretended—moral interests instead.

Other cases from the online half-world take an even more brutal approach to the secret private sphere, because word has got around that there is money to be made from exposing supposedly anonymous, romantic online activities. Several cases of 'ransomware' became known, which recorded users consuming pornography with the help of the webcam connected to the private computer and did not send this to the contacts from the computer address book only after payment of a handsome sum (cf. Pulliam-Moore 2015). In both cases, the fear that digital channels publish the supposedly secret love life is played with. The goings-on on the net, thought to be anonymous, are suddenly supposed to have an impact on completely normal offline life.

Of course, this crossing of boundaries also gives rise to an attraction, the attraction of the forbidden, which can only develop its power in conjunction with the possibility of being discovered. This motif seems to have accompanied love and making love in the digital realm since its earliest hours. At least this can be seen in Jonathan Franzen's novel *Innocence* in a flashback to the early days of the user-content-generated Internet:

> But even in the days of data-sharing protocols and alternative message boards, there was an inkling of the immeasurable dimension that would one day characterize the matured Internet and the social networks that emerged from it; in the uploaded images of someone's wife sitting naked on the toilet, the characteristic erasure of the distinction between public and private; in the aberrant crowd of wives sitting naked on the toilet in Mannheim, Lübeck, Rotterdam, Tampa, a foretaste of the dissolution of the individual in the mass (Franzen 2015).

Global, disappearing into the masses, but still as private as possible: naked, on the toilet, for the sexual edification of a foreign viewer. Even the most drastic representation of naked individuality becomes a faceless part of the mass in this digital context.

3.6 Advertising and Recommendations

The market economy produces an unmanageable number of products, differentiated down to the last detail. Take the product range of the online retailer amazon.de, for example. This consists of around 290 million products (as of 10/2016, including digital offers). In any given category, the selection is mind-bogglingly large: 30,940 baking pans, 18,643 electric gardening tools, 1067 products for air humidification and 949 products for air dehumidification. 966 types of needles and pincushions and on top of that 494 different bobblehead figures. The fact that this range of products can survive is a prime example of Say's theorem. According to this, every product supplied on the market creates its own demand. Without this effect, sales would not function in the highly differentiated, supply-oriented modern market economy. So it is central that the customers know about the supply. Communicating the existence of the product and making consumers aware of it becomes the central task: potential buyers need to know that the products exist and they need to know why they need them. It is unlikely that the prospective buyer of a bobblehead will take the time to click through nearly 500 variations. Nor will he want to order just any figurine at random. He has to rely on advertising and recommendation. We are now approaching this topic from two sides: On the one hand from the side of the company, on the other hand from the interpersonal and initially non-corporate side, specifically via the means of recommendation. For both areas, we trace three stages and show how they are intertwined in the course of digitalization.

3.6.1 Billboard, Newspaper Advertisement and Personal Recommendation

Times Square in New York is known for one thing above all: the colorful, flashing advertisements on the walls of the buildings surrounding it. This is where the country's major companies advertise their products. The city ordinance stipulates that the buildings must be covered by billboards in order to preserve the characteristic image of the square (cf. Hellman 1997) (Fig. 3.2).

Fig. 3.2 Times Square in New York City and its billboards. (Photo: Benoist 2012)

If the billboards were not digital LED walls, we would be dealing with a completely analog form of advertising. Which individual walker, tourist or employee of the nearby office buildings passes by the billboard, nobody knows. The reception of the billboard is not measured, much like another form of advertising: the classic advertisement in a newspaper. Let's take the example of an influential, national daily newspaper. The trade-off for their high reach is that the ad must appeal as generally as possible. The reader at the breakfast table may deal with the product information in detail, while the reader on the commuter train may casually skim the ad and routinely turn to the next article. Who ultimately is exposed to the ad, when and where, and with what consequence, is difficult to investigate. A short-term increase in sales figures after publication may be measurable, but it is difficult to link long-term effects to the ad. Only the reader sees the ad—and perhaps not even consciously.

Admittedly, the choice of newspaper can be used to adjust the target group. FliegenFischen, for example, the magazine with "current news from the world of exclusive fly fishing and the water guide" has such a clearly defined readership with a good 9400 bimonthly copies sold. A one-page advertisement costs a good 4000 €, and the quality and environmentally conscious fly fisherman is happy to be informed about the innovations on the market via such an advertisement (cf. Jahr Top Special Verlag 2015). However, if one wants to reach a larger target group with the advertisement, it becomes more difficult. For example, if one aims at "top decision-maker target groups" with a "comparably very high[n] level of their education", one should advertise in the Süddeutsche. The entire advertising page in the politics and business section, clearly where the decision-makers stop, costs around €84,000 there. At the weekend, when the decision-makers have more time to read their preferred paper, just under €100,000 is due for the full-page ad (cf. on this and the other data Süddeutsche Zeitung 2015). In the case of the BILD newspaper, the attentive reader and consumer is estimated more tangibly: "9.09 million readers own a motor vehicle: That is 80% of the readers". For a Germany-wide placement (11.34 million readers per issue) a proud €479,000 is due. In return, however, the newspaper promises its famous "unique reader proximity" (cf. BILD 2015). This spread between decision-makers and car owners leads to ads for either expensive watches or inexpensive cars. The respective addressee group served is nevertheless difficult to specify: It is simply assumed that, on the one hand, they are interested in a representative display of the time and, on the other hand, they are contemplating buying a new car.

The trade-off between reach and specialization, and the difficulty of measuring success, leads to another way to draw attention to a product.

Unfortunately, a method that is difficult for a company to buy: A personal recommendation. From movies to kitchen equipment to vacation destinations, a personal recommendation is the surest way to try new products. A recommendation comes closest to the personal experience and minimizes the risk of failure. To begin with, a recommendation is free of economic interests and therefore more trustworthy than an advertisement from a party directly involved in the sale. Therefore, it is the most credible indication of a product (cf. The Nielsen Company 2015). The personal recommendation is made in direct contact, when asked in conversation or when the use of the product is observed. Its quality is measured by the proximity of the actors involved: while no further research may seem necessary for a recommendation from a circle of family and friends, the more distant the sender of the recommendation, the more additional information or further recommendations become necessary. The possibility of recommendation is basically open beyond the circle of personal acquaintances and is then based on the reputation of the recommender. Test magazines, for example, benefit from this effect. If the magazine or the editor is a household name, this helps the credibility of the assessment. Stiftung Warentest in Germany or the Beobachter in Switzerland are simply trusted to do this.

In summary, the measures described are not unproblematic for the supplier side: billboards and newspaper advertisements are too far away from the customer and spread too widely. Accordingly, one does not get close enough to the customer and his private purchase decision. And a personal recommendation is difficult to force. Will this remain the case if the general development pushes back the analogue in advertising?

3.6.2 Quota Boxes and Direct Marketing

For television advertising, initially there is the same problem of dispersion: target groups such as "football fans" or "shopping fans" (beer or perfume advertising) and seasonal focuses (Christmas products) only narrow the groups very roughly, the dispersion is relatively large despite the high reach. A difference and advantage of television advertising compared to printed advertising is that it has a more direct effect on the customer with its combination of image and sound. The communication channels are doubled compared to purely visual advertising. Moving images create a higher level of attention (cf. publisuisse 2005, p. 9). How exactly, where exactly, however, remains unclear to advertisers.

The difficult measurability of advertising success is now being addressed scientifically. And it's being done with the ratings box. Market research institutes ask a representative number of households to install such a box on their television sets. It makes it statistically possible to show which viewers tune in or out for which programs (and commercials). Those who have a ratings box in their living room share what interests them, where their attention lies. By installing such devices in a representative number of households, it is possible to measure the number of viewers and how long they stay with a station. By means of extrapolation, the quota for the entire country is then determined.

The wall that has made the private sphere indeterminable and unknown to advertisers is cracking. The ratings box looks through the screen into the living room. The time when no one knew where the media consumer's attention was wandering is over. Thanks to moving images, sound and empiricism, advertising is moving a little closer to the customer.

The comparable development can also be seen for the area of recommendation. A model to make the recommendation systematically economically exploitable is called "direct marketing" and establishes an individual contact to the customer. The marketing takes place dialogically, the attention of the customer is high (cf. Dallmer 2015). The classic example is direct selling, which builds on direct marketing: Products are traded without going into retail sales, *Tupperware* relies on this principle. The sales partners, predominantly women, organise a *Tupper party* within their circle of friends and acquaintances. The products are presented in the premises of one of the female friends. After a tension-relieving glass of champagne, the presentation begins. The potential customers benefit from each other's experiences, one sees what the other buys. There is no obligation to buy, but who would leave their own friend out in the cold when she has already gone to all the trouble of organising the whole party. At the end an order list is circulated and the location of the next party is found out and who still knows interested girlfriends. Finally, the hostess receives a small gift. The principle of personal recommendation, the domestic-private setting and the group transfer the economic usability of personal recommendations to a sales method in a perfected choreography. The boundary between personal, non-economically motivated recommendation disappears in favour of an individualised shopping experience. The social environment becomes a carrier of economic function. Personal contacts become a monetizable network that is entered into lists and prepared for the next sales event.

3.6.3 Integrated Personalised Advertising: AdWorks and Spying Billboards

In the US feature film 'The Joneses' (2009) with David Duchovny and Demi Moore we get to know an interesting marketing operation: A pretend family settles in preferred residential area with the aim of stimulating consumption among fellow citizens. Each family member is supposed to present certain products as a trendsetter in their target group. In this way, the neighbours—out of interest, envy or social pressure—are to be encouraged to buy the products that are presented. 'Keeping up with the Joneses' stands in English—hence the film's title—for socio-economic comparison with one's fellow man. Implicit in this: the fear of getting the short end of the stick. The Joneses in the film play on this fear and reformulate it as an advertising strategy. The entire affluent neighborhood becomes a non-stop, unbranded *Tupperware party* through the covert courtship of the newly moved-in family: the Joneses carry, drive, eat, the neighbors do the same. People know each other, people want to impress.

The film exaggerates the game of need generation inherent in market-based product diversity. We recall *Say's theorem* mentioned at the beginning. It shows an advertising strategy that invades the personal sphere of the customer in order to get him excited about a product. With the test-tube family from the advertising agency's laboratory, advertising is suddenly no longer two-dimensional, no longer crammed between the pages of newspapers, no longer locked up in the television, but living next door. Advertising and customer are now separated only by the garden fence. How fortunate that these conditions are *Made in Hollywood*, you might think. In reality … we are already much further ahead!

In reality, digital networking also eliminates the garden fence that at least formally separated advertising and customers and the Jonseses. The blending of marketing, community, rating systems and the comprehensive collection of individual data enable marketing of previously unimagined precision. Dispersion is radically reduced, personalised. More relevant ads' and 'more relevant content' are generated at the price of the analysis of personal data. You only see ads for products you are actually interested in. These methods are already an integral part of current marketing campaigns.

For example, through the company *Clear Channel Outdoor*. It operates over 675,000 billboards worldwide and has managed to combine its billboards with behavioral tracking with its *spying billboards*—the company calls its product RADAR. The company installs devices in its digital billboards that register the unique identification numbers of cell phones that belong to people walking or driving by. The company partners with wireless carriers. Through data matching, consumer profiles can be created. As a result, content (based on knowledge of movement profiles) and the locations of advertisements are optimized. In combination with the retrieval of the same identification number at different points of sale, the company can additionally check the effectiveness of certain campaigns: The path from one billboard to the next to the point of sale and the purchase made become traceable. Although the company claims that personal identification is not possible via the collected data, data protectionists are fighting against these data-collecting billboards (cf. Nelson 2016).

With even less technical effort and even greater precision, the *Spying Billboard* also comes directly to the home. Smart TVs connect television with the Internet and thus enable the transmission of usage data to the manufacturer: what is watched and for how long, even how many viewers are in the room and what they are occupied with, becomes possible thanks to integrated microphones that are built in and continuously activated for voice control (cf. padeluun 2014). IP-TV providers such as Zattoo or Teleboy, which offer access to conventional TV programming via Internet streams, also process user behaviour: when, from where, for how long.

The data collected from scoreboards, smart TVs or homepages are valuable commodities. GoogleAdWorks is probably the best known platform of these goods. It offers companies highly accurate advertising. Ads can be shown on a location-specific basis or displayed only to users whose previous search queries show potential interest in the advertised products. The accuracy with which this information can be used to advertise in real time is shown by four examples from the current online world, which bring advertising unprecedentedly close to the customer:

- *Facebook.* When writing a status message—the usual short post on each user's own wall—the microphone of the input device can be activated. Facebook then analyzes what music or TV show is playing in the background while typing. The result can be automatically integrated into the status message. The addressees of the message can then purchase the music track or stream of the TV show via a quick link. A personal message on the pinboard thus becomes a purchase recommendation for Facebook friends in real time.
- *Spotify.* The music streaming service uses all the sensors and equipment available on mobile devices to find out which music is being listened to under which circumstances. Where is the listener (GPS), is he running,

walking, standing (gyroscope). pictures, contact data and media are evaluated (flash memory). Data is shared with partner companies to show more relevant ads and optimize response. Advertising and content are tailored to the customer—e.g. by creating a personal weekly playlist or adjusting the speed of music to the speed of movement (cf. Fox-Brewster 2015).
- *Pandora* is an internet radio station available in the US that selects and plays music based on the user's listening preferences. In the free version, Pandora fades in advertisements, matching advertising content with music. New, wild music on Saturday afternoon—adventurous advertising. Favorite band on Monday morning—conservative products. But that's not all: Pandora can determine a user's preference for a political party based on their preferred music genre, total tracks listened to, and favored bands. This has already been applied in various US election campaigns to display political ads to users. Based on the knowledge of the user's location, tailored to the corresponding constituency (cf. Singer 2014).
- *Amazon Echo*. *Echo* is a black cylinder of a good 20 cm height, which contains a Bluetooth speaker. But its real center is the high-precision microphone and the WLAN interface. The aim: *Echo* is to integrate the use of the Internet into everyday domestic life in a low-impact way. It responds to calls about the weather, game results and cinema schedules. Even the purchase of items is only a shout away. All requests are processed and stored on the Amazon server (cf. Tsukayama 2014). While Facebook, Spotify and Pandora involve digital collecting and linking, the *Echo* detaches itself from the digital space and materialises, standing as a product in the private space of its user.

Once the customer has been made aware of the relevant product (including political campaigns) in a targeted manner, as intended by the technical refinements shown, another means of selling that has become familiar through the online world has taken effect. We encounter again the principle of personal recommendation. Virtually all online retailers have set up a community system for customers to share their product experiences. The reduction of ratings to a point system (five stars have become established) brings clarity. It creates the impression of empirically proven objectivity, which shines through in the user reviews, unclouded by sales interests. The more reviews there are of a product, the more credible the overall assessment. A kind of 'swarm intelligence' on the quality of a product is created in the mind's eye of the prospective buyer sitting at home in front of the computer. One of the main problems of online shopping—the fact that the product cannot be examined live—is thus efficiently resolved. The de facto impersonal, unverifiable opinion of another customer is given the status of a personal recommendation. It is difficult to judge from the outside whether there are bought testers or even bots behind the reviews that influence the system to their advantage.

The different ways in which the various advertising methods get close to the potential customer can be seen in the development from the classic newspaper ad to the quota box and the dynamic online ad. Thanks to the combination of collected user behavior data and real-time analytics, there is now advertising that can adapt its multimedia content to the user's walking speed. The distance between advertising and the advertising target continues to shrink. The potential customer's need, whether expressed publicly or imagined quite privately, is identified and exploited in real time.

3.7 Surveillance

From the anonymous mass of employees of the American surveillance service NSA, one broke away in 2013. The then 29-year-old Edward Snowden made the global activities of the NSA and its Prism program public. He was the first to bring to light the extent to which modern government surveillance methods control citizens' digital lives. Potentially every email, video chat and homepage visit can be recorded, stored, searched for keywords, retrieved and processed (Lyon 2014).

Since doing without the devices used for wiretapping is not an option—a mobile phone is, after all, part of the basic equipment—one is largely at the mercy of such surveillance activities. Only the smallest possible acts of civil disobedience indicate a rebellion against this: the tape on the webcam of the laptop. The eye of the secret service should at least not be able to see into the bedroom.

The development of surveillance and its digitalization, which resulted in the helpless tape on the private computer, can be traced in stereotypical, medially mediated forms of the surveillant: At the beginning we imagine a secret agent, in a spy film of the 1950s. In black and white, he watches his target in the park through a hole in the newspaper. He has to be within earshot of his target, no more than a few arm's lengths away, to go about his business. We then the jump to the era of video surveillance. Here we see a detective feverishly analyzing images from surveillance cameras. In a small room of the police station he spends the night until, eureka, he suddenly catches the suspect on one of the many tapes. Then the cut to today. The investigator has changed. No more floppy hats, no more hours spent sifting through tapes. The *state of the art* in surveillance: a hacker typing cryptic lines into a terminal, gaining access to the smartphone of a gangster 1000 km away. One more input and the criminal's emails appear, one more input and we see him as a glowing dot on a digital map.

Away from entertainment stereotypes, let us take a closer look at these three stages of surveillance: from village policeman to video surveillance to blanket access to private communications.

3.7.1 The Village Policeman

He knows 'his friends' very well, knows where to look or ask when he wants to talk to them. The policeman enjoys a high personal standing and authority in the village community, knows where he can let mercy prevail over justice, where a personal admonition is sufficient. His work depends on his personal assessment of the situation, on his physical presence in the police station and on the street. Accordingly, he prefers to travel on foot or on a bicycle, equipped with the minimal insignia of his office, a hat, a uniform, he can gladly do without a weapon (cf. Gutknecht 2015).

The work of the village policeman is supported by the village eye. Jeremias Gotthelf first described this analogue surveillance instrument. At the beginning of the nineteenth century, the Swiss writer writes:

> From afar you could see that she knew they were looking at her, and that the eye of the village was open over her, when and how she went out into the field! O such a village eye is a good thing and keeps some in the egi! (Gotthelf 1838, p. 173).

The woman in Gotthelf's story sets off in the direction of the field with a display of diligence, because she feels the gaze of her fellow men on her. Whether by chance or by habit, the neighbour registers what is happening and sooner or later an exchange takes place about it. Awareness of this leads to the adjustment of behavior. This latent observation is a reason for many to prefer the city to the village—life is more anonymous when you only know your neighbour in passing. For our village policeman, however, the village eye is a helpful support in his work. Both through concrete observation and through the pure awareness that one cannot move unnoticed in public, the village eye standardizes behavior.

Taken to the extreme, we find this successful and traditional form of surveillance from the combination of village policeman and village eye in the settlement project *Celebration*. A town of just over 7000 inhabitants in

Florida, in the southeast of the USA. Celebration was planned on the drawing board in the early 1990s, and designed in a stylistically rigorous manner according to the aesthetic standards of the early nineteenth century, as envisioned by the nearby Disney Corporation. Including artificial snow at Christmas in the plaza in front of City Hall and aiming to be the safest city in the world. Anyone who buys a house must sign extensive contracts agreeing to abide by the rules: Cleanliness, friendliness and neighborliness included. Eager patrols are carried out, and tips from the population are immediately followed up by one of the many deputy sheriffs (cf. Eilert 2011). A view of the city from above shows how deeply the idea of mutual surveillance is anchored in the retort city: The layout of Celebration's downtown resembles Bentham's prison architecture. The major streets run in a fan shape toward a tall tower. Admittedly, this is not a tower topped by a guard. It is the 'Preview Center' from which newcomers can quickly get an idea of which part of the city they would like to live in (cf. MacCannell 2011, p. 31). The tower and the highly visible streets make it architecturally unmistakably clear to future residents that the basic layout of this city is grounded in openness, transparency—and control. Anyone who steps out of their house in *Celebration* is rid of their privacy. And they want it that way—neighbours greets you in a friendly manner and you greet them back, your movements are registered. In return, children can go to school alone and on foot.

3.7.2 Video Surveillance/CCTV

Video surveillance complements the village policeman and the village eye in one central point: humans are initially no longer necessary for surveillance. Instead of individual presence, the installation of a camera records impersonally and unfiltered what happens at a certain place at any time.

This surveillance is, let us think of a classical, fixed camera, locally limited, expandable by a higher number of cameras. Notices of recording are part of the strategy of this form of surveillance, the awareness of surveillance should lead to the reduction of undesirable behavior. Accordingly, the indications of surveillance often feature an eye depicted on a signal-coloured background—an updated and abstract form of the village eye.

The feeling of not being observed is annulled by surveillance. The knowledge of the technically conditioned storage of behaviour makes public space a space without privacy. Before, behaviour in public space was just as visible, but it was stored at most in the casual notice of another passer-by. By combining many thousands of camera recordings, every movement of every person who is in the monitored public space becomes reconstructable. The camera recording is stored, can be copied and passed on. The reproducibility of the situation is only limited by how long the recordings are kept. The camera becomes an iconic reference to the abolition of the secret private in public space. Either on legally mandated signs or, almost as frequently encountered, on critical engagements with the subject in public space. Like on this lamp-post in Basel (Fig. 3.3).

Clearly the robot-like camera construction is perceived as a threat, as a wired pistol on the chest of the filmed, the finger always on the trigger.

A special case of video surveillance and invasion of privacy is the so-called nude scanner. This technology is used to check large numbers of people at sensitive locations—typically airports—for weapons or other dangerous objects. In contrast to a sound-emitting detector, which beeps loudly, for example, when people pass through and after metal is detected—whereupon another person has to continue the search—the nude scanner displays an

Fig. 3.3 Threat and video surveillance. Own photograph clz, 01.06.2016, Basel-Stadt

image of the person being checked on a screen. On this, a clear image of the objects worn on the body emerges, as well as the sharp outlines of the rest of the body. Here, checks are carried out just under the skin—nothing remains hidden from the gaze of the naked scanner. Anyone who passes through here is defenselessly exposed to the gaze of the inspectors, at least for a moment.

3.7.3 Widespread Access to Private Communications: General Surveillance

Pub crawl? The child's first steps? Or a wild boar in the trash? The characteristic reach into the trouser pocket and the launch of the mobile phone camera. The definition of a novelty value in the true sense is not necessary to provoke a careful documentation of the situation. Everything is stored, documented, shared on the smartphone. And not just on the ground: Drones are conquering the skies and shooting never-before-seen footage. Recently, a man sunbathing on the narrow platform at the top of a wind turbine was filmed in this way and made the news (cf. Leuthold 2015). Even daring to climb a 60-meter tower was not enough to enjoy the secret-private pleasure of a few undisturbed minutes of sun with a view. Nowhere is safe from the prying eyes of a camera. Of course, the pictures end up promptly on the Internet, where they are viewed, discussed and disseminated globally. Every moment in every place is a potential memory, organized and accessible on all compatible devices.

This technical and cultural development sets the stage for a new quality of control, where every movement can potentially be shadowed via networked devices. Surveillance is no longer specific to individual population groups or certain personal, statistically generated characteristics (such as 'male, young, dark-skinned'), but it intercepts everything from everyone. Surveillance is general, so general that blameless citizens cover their webcam with tape when not in use—an expression of the constant awareness of control and invasion of privacy. Surveillance happens without the user noticing or even experiencing any restriction, remains entirely surreptitious and entirely painless. It takes place at night at home in front of the computer with the blinds drawn. It is very close to the human, but far from the user experience. The human factor still exists, at the beginning and the end of surveillance. But in between, there is a powerful data machine that obscures the view of the monitor. In the end, the search queries from the nightly surfing session may be read by an NSA employee—but when and where this happens and what this person looks like remains hidden.

Until now, the GDR was considered the standard of surveillance. Opening letters on a grand scale, bugging, tailing, inciting to report, IM neighbour and IM work colleague. But the real existing socialism in its German thoroughness is only the analog benchmark for spying and clean logging of the private. The difference to the *state of the art?* The Stasi bug in the apartment, the house phone, the letter did not accompany the monitored person everywhere. The constant locatability was connected with a huge personnel and cost effort. Today, surveillance is impersonal, universal, widely dispersed. Made possible by a whole orchestra of sensors at the disposal of the surveillants: Cameras, microphones, GPS receivers, motion sensors, constant networking with servers. Any cell phone can be used as a bug. A bug that, conveniently enough, is rarely more than a meter away from its target—after all, it's usually in your pocket. In addition, the metadata that has to be supplied by the mobile phone provider helps to create a precise movement profile. The data stream hardly ever stops. From the mobile phone call it goes to the computer, leaving traces with search engines and digital networks, the credit card number when shopping online and documenting interest in various topics by subscribing to newsletters. E-mails are intercepted on their way from the sender to the recipient, searched for keywords, en masse and without the need for an initial suspicion to be there. Denunciation is unnecessary with this dense information. The services only have to record.

Intelligence agencies are tapping into data lines on an unprecedented scale. The British GCHQ's "Karma Police" intelligence program targeted the intersections of the overseas cable between the U.S. and Cornwall, which carries about a quarter of the world's data traffic. The program allowed spies to log a profile of each user's browsing habits, as well as the exact user profile of individual homepages. Snowden revealed that the program logged over 1.1 trillion browser sessions by 2009—this was the beginning. In 2012, the program collected 50 billion entries on internet metadata. Per day. In the long term, every internet user should be provided with a unique ID and all the data of this ID should be stored in one place each time (cf. Watson 2015).

Modern surveillance is so successful because data-generating media are widely and universally used: Who doesn't write emails every day and own a mobile phone? Surveillance has moved so close to the observed that it can no longer just clarify undesired behaviour, but even thwart it. 'Predictive policing' is the name given to the method of using live data and existing police records to make predictions about likely locations, types and time of crimes (Seele 2017). The model is constantly updated and calculates—down to which block—the threat level (cf. Goode 2011). Based on prior events and knowledge of personal, socioeconomic factors, police could in the future use this method to make individual predictions of danger. Just divorced, lost his job, unusual cell phone activity, deviant internet surfing behavior? The program determines an increased likelihood of a crime. And made it clear that you are under surveillance—before a crime has even happened. You can hardly get any closer to a person's private life: you analyse a person's intentions and turn them against them.

Another example from the USA shows that digital state surveillance can also be operated with economic objectives. Analogue monitoring of road users is costly and personnel-intensive. This is changing with digitalization: In the USA, an Internet company offers individual counties the free installation of automatic license plate recognition software in police cars. The data generated is time and location stamped and stored on the provider company's servers. In addition, police are given access to another server at the vendor company where individuals with outstanding bills from court cases are stored along with their license plates. If the police vehicle then passes the vehicle of a defaulter, the police officer receives a message in a flash and can stop the person in question. The person then receives an offer: he or she will be arrested or can pay the outstanding costs immediately in cash or by credit card—in addition to a surcharge of 25%, which goes to the license plate recognition manufacturer (cf. Maass 2016). A mixing of state duties with private economic interests: Police officers are guided in their performance of duty by new motives related to the amortization of their own technical equipment.

Village policeman and village eye were surveillance instances with a personal face, which usually ended in front of a closed door of a house. What happened in the private sphere of the house was at most speculated about. With the widespread introduction of video surveillance, the control expands, becomes independent of time and personal effort. But even this surveillance remains external, enabling a private sphere to exist away from the lens. This changes with the area-wide surveillance of digital communication: access to webcams and e-mails, to search terms and mobile phone sensors allow the supervisors to participate in the life of the supervised in real time; a secluded, secret private sphere is hardly feasible under these circumstances.

3.8 Work and Employment

Working does not only mean producing goods and providing for one's own life. Social fashions, economic expectations and technical changes are reflected in the circumstances and fashionable forms of working. Digitisation has a fundamental impact on work in social, economic and technical terms. We trace the development of work through these three stages and also take a look at the private sphere of the working person: At an agrarian, nature-based work rhythm, through industrial work with a time clock and, the most current variant, the smartphone tracking of the employee by their boss.

3.8.1 Natural Working Rhythm

The cow has to be milked at five in the morning, so the desire to sleep in is of little help. The rhythm of the cow dictates the rhythm of the milking. The same applies to farming, which is based on natural cycles and not on holiday or work schedules. Working with animals and plants illustrates what a natural rhythm of work means and sheds light on pre-modern working conditions that were oriented towards the change of day and night and the seasons. Work took place near the place of residence, often the place of work was identical to it. Due to the lack of technical possibilities—there was no electricity and technical equipment for harvesting and tilling the fields, but a small-scale transport infrastructure—work was restricted to a comparatively small radius of action. The object of work was oriented to this framework and thus consisted primarily of simple agriculture, subsidiarily oriented to the workers and their basic needs. The natural conditions and necessities shaped the work and the economy related to the individual household and the body-owner. Any surpluses could be sold and reinvested.

The power of disposal over labor and its fruits varied in the medieval social structures. Free peasants organized themselves independently in extended families or farm associations, semi-free peasants likewise, but without social rights. They could be called upon by higher classes to perform services. Unfree and serf peasants, who worked for landowners and nobles, in turn had fewer rights. As a consequence, self-determination over one's own work, its organization, and the treatment of its results declines. The worker benefits from it rather indirectly and abstracted from the result—exemplified by the reformer Martin Luther, who ascribes a religious function to work apart from production and acquisition (cf. Zapf 2014).

The free peasant of the Middle Ages probably embodies work in the rhythm of nature best. Within the small scope of action of the technical possibilities available to him, he works for the subsistence of his family. Any small surpluses are reinvested in this process, but without forcing the growth of the farm. This expansion would not be easy: without a social or technical cooperative infrastructure, the work is entirely related to the farmer himself, his field and his livestock. It is, in business terms, non-scalable labor, limited to the physical labor of the farm owner. Work and non-work are close in space and time. The farm is a place of residence and work, and free time is arranged along the requirements of farming and the needs of the animals. The break in the barn, the nap in the woods, of which the other family members are unaware, is part of it. The results of the work vary: Fluctuations in yields are largely due to higher forces. And even in good years, the status-subsidiary oriented work is not crowned with riches. But neither is it geared to the immediate re-investment of the small profits made to enlarge the business, as would naturally be good practice in the capitalist economic order (cf. Weber 1986, p. 53). Labour fulfils the purpose of meeting the direct needs of those involved in it and generating enough to pay taxes. The means and ends of work are closely linked to the worker himself.

Example: In James John Hill's 1851 painting *Shepherd Boy and Shepherd Dog*, we see a young shepherd asleep with his tools in his arms, his dog also asleep beside him. In the lonely, hilly landscape in the background, one thinks one can make out the grazing sheep, the sea looms in the distance. No external influence affects the dozing shepherd. The sleeping dog, normally frantically busy rounding up the sheep, reinforces the restful, unobserved overall impression. The romantic, transfiguring vision of the painting pushes the shepherd's work into the background, emphasizing the connection to nature, the tranquility and undisturbed nature of his craft.

3.8.2 The Time Clock

Mechanization and industrialization allow productivity to rise to unimagined heights. The only limiting factor is the human factor, and so this is optimised: Taylorism starts its efficiency increase with the worker. With a powerful stamp on his time card, the worker enters a working world in which every single step he takes is monitored. A hierarchical system of superiors, division of responsibilities based on the division of labor, and monitoring of quality, quantity, and target fulfillment.

Fordism, after Henry Ford, the world-famous US car manufacturer, is also based on this work model. Ford perfects assembly line work, makes full use of the technical possibilities, controls processes, lowers prices and thus increases sales. When it's time to punch out in the evening, the whole optimization spook is over. If not too tired, the successful assembly line worker can drive his own car to a well-equipped home and recover undisturbed from the stresses and strains of the working day.

With the transition from an industrial to a service society, this applies not only to the assembly line worker, but also to the office worker. Imagine a man, very much in the style of the TV series "Mad Men". He lives in a New York suburb in the 1950s. Day in and day out, he takes the same train to and from work. During office hours, the man is a professional, working to strict specifications, reporting, optimizing, controlling. When he boards the train in the evening and comes home, closes the front door behind him and greets his family, he is a private person. Duty is duty and schnapps is schnapps.

Both forms of work—industrial work and work in the early service society—are the radical counter-design to subsidiary-agrarian work. The objects of labor break away from their natural limitations and multiply, adapting to the various requirements of diverse products and offerings on the market. This economic expansion fundamentally changes the circumstances of pre-industrial creation: the unity of the place of work and the place of residence is abolished, working time and leisure time are more strictly separated. Working time is no longer dependent on nature, the link to natural rhythms is largely removed. Work is done indoors, lit and heated. Industrial work has its own rhythm with an organisation based on the division of labour, a complex supporting infrastructure and the expansion of the economic radius of action.

3.8.3 Smartphone Tracking by the Boss

Equipped with his laptop and a cup of coffee that is oversized by European standards, the modern city dweller takes a seat. And starts working while still at the coffee house table.

This is an expression of unbounded working conditions: the detachment from the field, stable or office as clearly defined workplaces. Starbucks, the large US coffee house chain, has even incorporated this practice into its corporate philosophy. The store is supposed to be a "Third Place", a place that is "the third most important place in people's lives, next to the home and the office" (Starbucks 2014, p. 2). An in-between world, then, between work and leisure, with high-speed wi-fi, operated by a global gastro-franchise. The separation between work and non-work becomes local and temporal, flexible.

Less structured areas of work—for example, those of the writer or the scientist—have always not been bound to fixed working locations or hours. The expansion of flexible forms of work beyond these few professions stems from the possibilities of digitalisation. Mobile phones and e-mails make it possible to establish contact outside of conventional office infrastructures, mobile computers and networking make the rigid restriction to work location and office hours unnecessary. This initially means more freedom for the worker, releasing him from his fixation on landline telephones, filing cabinets and postal addresses.

This potentially disjointed work is now being supplemented by digital networking: data-supported HR work creates a profile from application documents, personnel files and digital self-disclosures (e.g. on Facebook). The fit of the applicant for the position to be filled or the employee with his tasks becomes mathematically verifiable. Factors such as social activity, leisure time behavior, the length of time spent at a location or workplace, education, marital status, even behavior and statements in social media are included in this evaluation—and have an influence on hiring or promotion. The employee's desire to change or career planning becomes visible to the company, even if the employee has not yet concretized this for himself or even communicated it (cf. Seiwert 2015).

As a rule, the employee is unaware or unaware of the data collection and monitoring. It happens in the background, running along with everyday tasks such as writing emails or visiting websites. Everything is digitally logged. The workflow is not disrupted by digital monitoring and logging. Unlike a foreman who loudly urges workers to be diligent every ten minutes, the electronic sensorium is invisible. The control only becomes real when—evaluated and compared with the data of colleagues and the demands of management—it is printed out for the quarterly discussion and has found its way into the personnel file. The collection of information is technically easy—for example, location data in a logistics company. Here, the boss has a vital interest in knowing where his employees are at all times: Are they delivering packages in the optimal order? Are they perhaps spending too much time chatting with customers as they drop off the parcels? Or are they sitting on the edge of the forest in their van, taking a midday nap? Thanks to an app on the drivers' service smartphones, the employees' location is continuously sent to the boss's computer. Who, where, how fast, how long—all this is recorded and stored. Over time, the systematic collection and analysis of employee data creates a comprehensive picture. This information is not only used in the interest of the employees, as is the case with the retail giant Amazon in the United States. The company defended itself against a former employee who publicly criticized the company by disclosing confidential details from the employment relationship—for example, that the critic cried at work on several occasions as well as the critical assessments of the supervisor before a promotion. This information from everyday work life reduced the credibility of the critic in the reporting, the criticism was made small with this public disclosure of employee data (cf. Person 2015).

The logged and stored browser history, the access to company emails, the tachograph from the company car, the logging of the state of mind at the desk. By signing the employment contract, the employee must agree to this data collection. According to current case law, this monitoring practice is legal during working hours. It becomes problematic when the surveillance creeps into the private life of the employee and the smartphone also sends the employee's position in the evening and on weekends (cf. Knibbs 2015). Admittedly, the data is rarely evaluated—usually only when something serious has happened. Nevertheless, the awareness of surveillance is enough to normalize the behavior of employees. The eye of the boss is always on the worker. The resulting norming extends well beyond the work environment into leisure time. Systematically, the boundary between work and non-work is erased: Google, for example, offers so many amenities on its premises that individual employees move onto the premises altogether. From showers to restaurants and shopping, the company aims to give employees the status of family. With this in mind, Apple and Facebook recently began offering to pay for egg freezing for their female employees. In the best years, power can thus be used for the company, and family planning can be postponed until later in agreement with the company (cf. Wisdorff 2014). But the dissolution between work and private life is also evident in more subtle processes: Within companies, people are connected on social networks—know about each other's leisure activities through it. Sipping sangria from a plastic bucket with a bare-red torso (and posting that), even if it is undoubtedly a leisure activity, thus takes on a work-relevant character. The "normative expectations of professional respectability of employees [is] transferred to their private Internet activities" (Crueger 2013, p. 22).

From the Middle Ages and unsupervised work in nature to the supervised office job today, productivity and prosperity have risen steeply. And they rise, it seems, with each additional control and innovation of surveillance that is introduced. At the price that the unobserved goatherd snoozing in the forest clearing now belongs on a farm holiday. With the increasing digitalization of the work environment and work objects, the secret private disappears from work.

3.9 Election and Political Advertising

The political election is the crystallization point of a civic secret private. In a constitutional state, the secret is a constituent, constitutionally protected part of the electoral process: elections are free, general, equal—and secret. In the following, we trace the development that digitalization has brought about in voting, especially with regard to the secret coupled with the election. Connected to this—and subject to the same change—is also the courting of the voter's favour, i.e. the attempt to externally influence the privately formed and secretly expressed electoral decision of the citizen.

3.9.1 The Secret Ballot and the Election Poster

The elections to the People's Chamber of the GDR knew no secrets. Voting booths were provided *pro forma*, thus theoretically enabling the secret ballot. In practice, however, since there was only one list to choose from anyway, most people only received the ballot paper and openly signalled their agreement with the list by folding it and dropping it into the ballot box in front of the bystanders. Election results with over 99% approval were thus no surprise and simulated the unity of party and population (cf. Baum 2002). Another example of a non-secret election in the democratic tradition is the Swiss Landsgemeinde. Here, all citizens entitled to vote meet in a publicly accessible place. There, votes are taken in person and in public on the business on the agenda. Majorities are determined by looking at the crowd (cf. Stadler 2008). A type of voting that is structurally not secret, the sideways glance to the person next to you cannot be avoided. For this reason, and for pragmatic reasons—the number of eligible voters has now become too large—the Landsgemeinde is now only practised in two cantons (Glarus and Appenzell Innerrhoden) and has otherwise been replaced by the ballot box.

These two examples demonstrate the value of the secret ballot: it contributes to the fact that electoral decisions can be made without the fear of reprisals—be they state induced (GDR) or socially induced (Switzerland)—and thus more freely. Finally, the election result should paint a realistic picture of the population's mood, and accordingly the basic assumption is the following: the awareness of the secret ballot makes a free election possible.

It is part of a democratic constitutional state that political conditions develop in a battle of opinions, not only in parliament but also in private. Standing up for one's own convictions, at the same time listening to the opinions of others and examining one's own positions in exchange. At this point in the formation of opinion, the legally guaranteed secret private sphere becomes a piece of the public sphere, and yet: the point at which the cross is finally placed and then neatly folded and lowered into the ballot box remains secret. At the moment of voting, there is by definition only the voter's own knowledge, nothing else should influence him. The electoral decision as the product of a secret-private conviction, which materializes as a cross in the voting booth at the moment of casting the vote, and then unfolds its effect anonymously, in conjunction with the electoral decisions of the other citizens. This process materialises even more consistently in the postal vote, the secret ballot in private. Again, using the example of Switzerland: the election documents are always sent to the homes of the eligible voters. The mailings always include the voting documents as well as information material such as the positions of the Federal Council and the pro and con arguments of political parties. Around 75% of voters subsequently vote by mail (cf. Villiger 2014). In Germany, on the other hand, the prevailing view is that the election represents a "public sphere based on the rule of law", which ensures both the transparency and legality of the procedure and a visible act of civic participation. Although postal voting is no longer subject to justification, it represents a significantly lower proportion of voter turnout than in Switzerland, at less than 20% (cf. Gröschner 2013).

The secret vote cannot be traced back personally, neither in the case of secret voting in the voting booth nor in the case of secret-private postal voting. Accordingly, election advertising in the legal state proceeds via the secret private of those entitled to vote: advertising has to persuade individuals and thus attempt to influence personal decision-making. The tightrope walk between promoting a reflective decision and influencing it is thus inherent in election advertising. It is regulated accordingly: it must be labelled as such. It must also be open to smaller parties, and it must not take place in the direct vicinity of the election. Its financing must be transparent. The classic form of this election advertising is the presence in public space, through posters and public appearances, combined with the tour of the candidates: Sticking posters, being personally approachable in public, door-to-door canvassing and trying to convince voters through direct contact. A common form of opposition to this kind of election advertising in public space is the smearing or tearing down of election posters and the display of flyers. Both variants use the anonymity of analogue forms of advertising to demonstrate their opposition.

3.9.2 Voting Machines and Civil Dialogue

For the sake of the secrecy of the ballot, the medium on which the vote is cast seems to have a special significance. Otherwise it is difficult to explain why the classic paper ballot is still so widespread despite full-scale digitalisation. The introduction of electronic means in voting is proceeding hesitantly—not least because of negative experiences such as those in the US presidential election of 2004, where irregularities occurred in the state of Florida in connection with touch-screen voting machines that turned out to give a systematic advantage to the candidate George Bush (cf. Ziegler 2004). Not that an election held on paper lends itself less to manipulation. But it seems as if the medium has an influence on the election. There is no other explanation for the fact that voting behaviour in the Swiss parliament changed after the installation of an electronic voting system: with the system, which makes the results public in real time and thus enables direct comparability, there is less unanimity in the Council and at the same time more unity within the larger parties (cf. Bütler 2015). The changeover of the voting medium on a large scale is thus less a question of technical possibilities than one of cultural consent: Only pen and paper are trusted to actually fulfil the promise of a free and secret election.

The situation is different in election advertising. This shows itself to be open to communication with new methods and media. As early as the 1969 federal election it is reported that there, for the first time, all three major parties (CDU, SPD, FDP) used advertising agencies, which had previously only advertised consumer products, as advertising experts. Now the focus is on mass interaction with the voter: Televised question-and-answer symposia, the deliberate promotion of people, characters, minds, in conjunction with blanket advertising on radio and TV was intended to open up dialogue with voters. The election becomes a product that moves masses closer to the voter than was previously possible (cf. SPIEGEL 1969). The communicative one-way street 'election poster' is supplemented by dialogical measures and the inclusion of new media.

3.9.3 Obama and Pandora

In line with the observation that the electoral process itself generally takes place conservatively with pen and paper, the effects of digitization are most apparent in election advertising.

The digital is changing political decision-making: People who don't always vote for the same party out of habit previously had to do extensive research and spend hours comparatively reading party platforms to form their opinions on how to vote. Now that can be done in a matter of minutes: Those who don't know what to vote for can get an online evaluation of which party program best fits their chosen answers by filling out a questionnaire. Wahl-O-Mat or Smartvote know what you want to vote for. In addition to providing neutral information to the voter, such digital infrastructures are used to evaluate voting behavior for partisan and economic purposes. For example, the American music streaming provider Pandora has succeeded in developing an algorithm that can predict users' voting behaviour on the basis of their individual musical tastes and listening habits (cf. Singer 2014). This insight is used by the company to make money by targeting the accounts with the known voting preference with targeted election ads. Pandora offers politicians in contested districts to specifically consolidate or attack individual voting decisions.

The data-driven analysis of digital usage behaviour with the aim of predicting election decisions has become so elaborate that it is seen as a key method of Barack Obama's election victories in 2008 and 2012. Obama relied fully on the data-driven organization of his campaign. In this way, he was able to convince voters precisely in individual, undecided parts of the country—down to the level of individual city districts—even in politically fragmented areas, where the resources of a traditional campaign would not have been sufficient or would have been too ineffective (cf. Balz 2013). And much less elaborate methods are all that is needed to leverage digital networking for electoral success. One research suggests that personal voting recommendation on social networks significantly influences network voting behavior: A voting message from friends encourages people to vote themselves (cf. Lobe 2015).

Digital campaigning methods continue to expand with two goals: To map changes in decision-making in real time and to communicate directly with voters. With the knowledge of current search queries and page views—currently still reserved for the operators of search engines—a real-time map of the efficiency of election campaigns could be created, broken down to individual reactions regarding individual information. The same data offers the possibility of directly contacting individuals currently engaged with the issue in order to give the opinion-forming process an appropriate spin. Or, more subtly, to link results on individual search terms favorably to particular candidates (cf. Rogers 2015). Accessing search data and correlating it with individual voters would further increase campaign efficiency.

So, despite ultimately voting with pen and paper, algorithms make it easy to accurately predict an individual's voting decision based on usage behavior with search terms, texts read, videos watched, and cross-referencing with musical tastes and other indicators. Moreover, unlike a pulled-down poster or flyer, digital, political expressions are easily traceable. The operators of social networks and search engines gain unfiltered access to the political opinions of their users. This is an asset of which the companies are aware and, see Pandora, use extensively. Thus, every action on the net, every video clicked or product purchased online becomes a potential contribution to the reconstruction of voting behavior. The information becomes a monetary benefit for selling.

With the digital, the secrecy of the ballot is eroding and being replaced by a mixture of economic and political influence. The private sphere is now truly political, albeit somewhat differently than the 68ers imagined. The private sphere is now being used, the secrecy within it removed, in order to get closer to the needs of the citizen and the voter.

3.10 Networks

A 'network' first generally means an infrastructure that connects individual points—for example, with the aim of exchanging energy, goods or information. In the form of energy networks such as electricity grids, transport networks such as roads or information networks such as telephones, they are a common part of everyday life. Digitization has had a particularly strong influence on information networks and, with the Internet, has fundamentally reformed communication and information channels (cf. Newman 2010, p. 18).

For our topic of the secret private, another type of network is interesting: *social networks,* today synonymous with online platforms such as Facebook or Twitter. In its sociological meaning, the term does not only refer to the digital version of the network. It refers more generally to a loose or institutionalized organization of individuals under specific goals and to varying degrees. Separate networks are found for different social interactions. Unlike their material counterparts, which structure the exchange of energy, goods, or information, these networks emerge only through social exchange. They are immaterial networks that make use of other networks for their functioning, first and foremost those of information exchange.

The common thread for the following presentations is: digitized social and informational networks are changing their content. We will look at how exactly, in a three-step process, starting from the bulletin board and the connected analogue-personal network to crowdfunding and today's powerful, unbounded digital networks.

3.10.1 Pinboard

Analog social networks are primarily based on personal contact. These are the family, the closest friends, the work colleagues and the direct contact with each other in this context. Here, information flows in conversation, in informal personal exchanges. In terms of economic interests, such networks are referred to as rope networks, and their consolidation is critically referred to as *nepotism:* People know each other, people help each other. These networks thrive on the fact that they are not transparent to the outside world and information is only disclosed selectively.

Such a loosely structured network for the exchange of information and mutual benefit can also be described outside the direct, personal environment in the context of an institutional affiliation. Such networks, which result from the common membership of certain groups and serve the exchange of information, are described in research as *affiliation networks* (cf. Newman 2010, p. 53). A simple example of this is *the notice board,* the bulletin board in university or school. A designated place where members of the institution regularly stop by. The advertisements are appropriately themed: Free shared rooms, used bicycles, event announcements. The carefully prepared, tear-off strip with the telephone number is the only indication of how many interested parties have probably taken note of the information—otherwise the network is a trackless affair. Professionals recommend tearing off the first strip as soon as it is hung up, because this signals demand to the next viewer. The information can only be found on the spot by physically going to the pinboard.

Another example is the newspaper advertisement for finding an apartment. Here the loose network of newspaper readers is used and here too there is initially no feedback between the recipients and the originator of the communication. This restrictive and initially traceless communication is further intensified by the use of a cipher number, behind which the provider disappears and can only be contacted via the newspaper publisher. Communication only takes place if there is mutual interest.

Pinboard and cipher display are part of social networks, which are characterized by a high degree of non-commitment and low personal traceability. Accordingly, the dispersion of these networks is large. On the other hand, the entry barriers are low. The network used to transmit information is analog—on paper—and tied to a personal contact with the information: Passing the bulletin board, buying the newspaper. This analogue information transmission results in uncertainty about the success of the corresponding communication—there are only uncertain clues, such as the number of torn strips or the circulation of the newspaper.

3.10.2 Analogue-Digital Information Networks

Uncertainties about reception and the lack of efficiency of analogue information networks led to the rapid adoption of digital infrastructures in the areas described. The quality of informational aggregation could be drastically increased as a result: Time- and location-independent access improved the findability and number of views. Accordingly, formats that previously relied on the personal or *affiliation network* are now found in digital information networks. Classified ads, forums for tracking down former classmates, or a computerized search of libraries are examples where the information network has changed without changing the subject of the particular search. The Internet auction house eBay stands as a successful model for this transfer. The bargain flair of a flea market with flexible prices, transferred to a digital platform. The digital infrastructure turns the clearing out of the old children's room into a sales event, only without the early rising and tedious setting up of the sales stand. Both worlds, the analogue and the digital, come together here.

Economic networks are also changing along with digital networks, expanding into the private sphere. Deutsche Telekom was the first to do this when it went public in 1996. Not so easily: the investors were to be broadly spread among the population, not just abstract investment funds, but a *people's share was* to capitalize the company on the market. With great media fanfare, the acquisition of individual shares by private individuals was praised as a novel investment strategy. The once exotic world of stockbrokers—a conspiratorial community with secret hand signals and opaque strategies—was opened up in the media, and participation was marketed as an opportunity for ordinary citizens to take part in the great economic events. Most small shareholders had a relationship with Deutsche Telekom at the time they bought their shares, at least in the form of a simple telephone conversation, an image of the company as a solid former state-owned enterprise. They were lured by the promise of good profits at low risk. Easy money. The information was well received, newspapers and television were full of it. Then came the year 2000, and with it the reality of stock market investments also took hold of the people's share. The paper suffered massive losses, of the more than six million individual shareholders, around three and a half million remained after the price fluctuations—the rest sold their shares and withdrew from the financial market (cf. Hagelüken and Jalsovec 2011). The world of brokers and bankers remained a brief excursion for most. More traditional investment methods were quickly preferred again. Nevertheless, the T-share paved the way in bringing the world of finance a little closer to the general population, even if today it is no longer in the form of individual shares. Today, retail investors are a customer group to be taken seriously by banks, not least because they can obtain information about global stock market events in real time and virtually free of charge, enabling them to make independent, informed investment decisions. Through the digital network, the PC user ecomes a trader.

A further development of these small investor activities is *crowdfunding*, a part of the sharing economy. Through online platforms, people present projects that they need investors to realize. Individuals can support the projects with small amounts, either as a donation or in conjunction with discounts on the future product. Mass participation makes it effortless to raise larger sums. The digital platform takes over the function of the intermediary. The risk for individual investors is manageable due to the usually small individual investments. In addition, the emergence of a community enables a monitoring effect that functions similarly to a personal network with reputation effects and deters free riders (cf. Hui et al. 2014). Novel digital network opportunities make it possible to become entrepreneurial through social networks.

3.10.3 The Powerful Digital Network

In the digital network, all kinds of information is available practically free of charge and in large quantities. This necessarily leads to a change in the search and retrieval of information, in short: to an increase in the importance of infrastructure. This is particularly evident in the role of search engines, which organize knowledge and enable access to the seemingly endless and confusing global pool of knowledge. The ordering of the world by the search engine takes place under a certain idea of relevance: search engines rank results and thus make an evaluation. The experienced surfer knows: The search results on the second page are practically already invisible. The rankings are generated in order to fit with the search term and, here the evaluation comes into play, according to the popularity of the page. The search result therefore says nothing about the quality of the content. It is not possible, based on the search results, "to distinguish between patrician insights and plebeian gossip. The separation of high and low, the serious and the trivial, and their intermingling at times of carnival date from times past" (Lovink 2011, p. 1). Thus, the information network influences its objects and its users via the need for structuring and is thus a normative endeavor (cf. Stalder and Mayer 2011) For example, Google suggests the addition of search terms and even substitutes, when other terms are entered than those the search algorithm assumes, the terms expected instead via the actual search query: did *you mean* … and what is meant is displayed. Only with an additional click can the actual query be displayed. Google knows better than the searcher what is being searched for. 'I say one thing, but I mean another'—and Google knows which is which. That's how it has to be in a good relationship.

In this web of order, relevance and opinion, search engine providers have a powerful position. There are no specifications for the organization of the search results. Manipulations would be correspondingly easy to create. In the case of elections, it would be possible to link candidates with certain search queries. Already in use is a mechanism that parries search queries that indicate criminal interest with behaviour-changing counter-suggestions. Anyone Googling for ways to join a terrorist group (e.g. 'join IS') is supposed to be shown offers that discourage them from doing so. The company sees this mechanism as part of an effort to generate counter-narratives to the terrorists' stories and strategies and their recruitment of followers. Google works with relevant NGOs to offer these institutions free premium placements on ads displayed along with search results (see Barrett 2016). Already in the attempt to use the network, the network gives its own direction to the cause. In the case described, admittedly, for the better. Other applications not excluded.

The changes triggered by the digital network can also be seen in more everyday examples, such as the search for an apartment or a job. In apartment hunting, the digital information network is leading to a reinterpretation of the role between tenant and landlord. The Smmove platform, for example, has made it its mission to transfer the entire process to the digital realm. Prospective tenants create a profile and use it to try to convince landlords of their suitability. The more information and effort that goes into the applicant dossier, the higher the chances of getting views, perhaps even getting accepted. Once the landlord has selected a group of applicants, they can bid for the apartment. As in an auction, the prospective tenant who offers the most rent wins the bid (cf. Fabricius 2015). In the market for real estate in sought-after cities, the customer must first offer himself to his provider and then hope to win the bid through successively higher prices. The digital network makes this possible.

A similar change in negotiating position can be observed in the job search. Within platforms such as LinkedIn or XING, applicants provide information about themselves in the hope of being discovered by a passing headhunter. If someone is interested in an applicant, they are screened through the same network prior to possible hiring by gathering the available information and reviewing it for compatibility with the employer. The potential applicant does the same thing when researching the company's name and looking through reviews before applying.

Getting to know each other in person is only a first step within these networks and not absolutely necessary. The term 'friends' refers to the status of a digital connection, of which each user of the market leader Facebook can boast an average of around 340 (cf. Beeger 2014). In the search for new friends, an algorithm is helpful, which displays with amazing precision names that are actually somehow known: Was also at a party, know from the canteen, prefer the same café. Through the generous inclusion of friends and the regular and extensive use of one's own profile page, a selective picture of one's own life is drawn. With a strange consequence: the reception of self-representations by others is perceived as depressing and unsatisfactory (cf. Dribbusch 2013). So many parties, holidays, cars. You yourself sit at home and click through the exciting lives of others.

Another change in the digital social network is that it can be easily used for external purposes due to the combinability of digital information. For example, one revenue stream from Facebook is the sale of targeted ads. Companies can track individual users and display advertisements to them across applications—thus advertising concepts can be realized that accompany the user in various digital applications (cf. Facebook 2016). According to the user profile, not only the advertising content but also the information of the network itself is customized: Messages from closer friends are displayed further up, frequently searched information is placed more prominently. The end user moves in an informational space tailored to him or herself, which is optimized to attract as much attention and interest as possible—the Internet user finds himself in a filter bubble without necessarily realizing it (cf. Pariser 2011). The fact that digital social networks can be used for political manipulation due to these possibilities of individual adaptation is made apparent by the example of Weibo, the Chinese micro-blogging service. There, the attempt of some user groups to emphasize political content more strongly was registered. In order to counter this development—because the scattered information can only be controlled with great effort—a layout was developed that is more image-heavy. Content for entertainment receives above-average visual attention compared to text, while the communication of more complex content is pushed into the background. As a result, and following further interventions of this kind, the Internet in China, along the official concern of uncontrollability, is now largely perceived as an economic rather than political space. The censorship of the Internet as a process with an economically positive effect while at the same time suppressing organized criticism of the state leadership (cf. Benney 2013).

One problem with using digital social networks is creating trust and commitment. Here, too, China has come up with an innovative solution: the credit score. This has an influence on whether loans are approved and is made up of the analysis of purchasing and payment behavior and also behavior in digital social networks. With numerous online friends and an uncritical external appearance, the score improves. With the wrong friends and critical comments, it goes down. The score is supposed to be publicly visible—and thus functions at the same time as economic and social leverage to normalize behavior (cf. Falkvinge 2015). Here, social networks, political and economic interests, and the private expressions of citizens merge into a single number.

These examples show how the digitized infrastructure allows different networks to merge by collecting, combining and processing information. Social, communicative, political or economic networks fed with private information move together into a single digital user experience. The omnipresent digital network is moving away from its original purpose—the informational connection of individual points—to become an independent, influential entity in which there is no longer any secret.

3.11 Payment and Digital Currencies

The everyday process of payment is being transformed by digitalisation and in the process directly touches on our question of secrecy: cash flows reveal a lot about their originators. When customers pay even small amounts without cash, accurate conclusions can be drawn about their consumption habits, eating and travel habits, financial situation, interests and social activities. The more digital the circumstances of the purchase, the more difficult anonymous payment becomes. An immense data trail is created. In addition to the usual data generated by website visits, it is credit card data, data from payment service providers such as PayPal or traditional banks that make the purchase transparent. The fast, anonymous cash payment has stopped existing in the analogue space. Starting with the credit card and ending with the purely digital cryptocurrency, a piece of secrecy is lost with every further digitization of payment.

3.11.1 Cash

If you do not want your purchase to leave a trace, you pay cash, of course: I don't need the receipt, thanks. Be it for an innocent surprise gift that should not appear on the bank statement. Or at the relevant shop near the station.

The purchase leaves no trace, no name, no address. The cash payment is anonymous and analog.

Even larger purchases can be made this way, at least in Switzerland. The 1000 note there is one of the most valuable in the world. An achievement of anonymous payment that can no longer be taken for granted. The 500-euro note was abolished in 2019, which is intended to counteract illicit work and terrorist financing. The measure seems to satisfy mainly a signal effect, as empirical evidence of a decline in crime after the abolition of large notes has yet to be found (cf. Knupfer 2019). Admittedly, dealers are reluctant to accept a credit card and tracking cash involves a lot of effort. Keyword: 'numbered notes'. This seems to be the price of cash. And this is predominantly legal: in German-speaking countries, over 70% of all spontaneous transactions are carried out in cash (cf. Bose and Mellado 2019; Bruckmann and Eschelbach 2018; Zulauf 2018).[4]

Storing money in cash is also low-trace and anonymous. In times of the international automatic exchange of information between banks and tax authorities, a traditional combination is becoming topical again: bank safe deposit boxes are experiencing increasing demand (cf. Grundlehner 2018). What is put into the safe deposit box is not recorded by the bank. And 100,000 francs in 1000 franc bank notes are only 1 cm high.

For a long time, the simplicity of cash payment was the trump card and guarantor for its continued existence. It is time-consuming to insert the card into the reader and then to enter a PIN or signature. The trained cashier is much faster with cash payments—often he has the right change in his hand before the note has even been taken out of the wallet, as the customer knows. Thanks to new card payment options, however, the advantage of speed counts for less and less. Contactless payment and billing via mobile phone increase acceptance and practicability. The digitalization of payment is advancing and reducing secrecy when paying.

3.11.2 Credit Card

The success story of the credit card can be traced back to two factors. The convenience of cashless payment. And the possibility of obtaining low-threshold short-term credit for consumer spending. The credit is fixed in

---

[4] From an economic perspective, however, things look different: In the euro, cash accounts for only about 10% of the total money supply. Another 10% are demand deposits of the banks at the central banks, and a full 80% of the money supply is fiat money, which is created by banks through lending and does not function like cash (cf. Schäffler 2019).

advance and no longer has to be applied for individually—with the credit card, taking on debt "suddenly becomes convenient because it is impersonal." (Pitzke 1999, p. 4). The issuer of the card knows about the financial situation of the cardholder and finds out where and on what the money is spent. In addition to interest on the credit granted, this generates valuable information for the issuer about the purchasing behaviour of its card customers. The terms and conditions of credit card companies reflect the dual business model of credit and data collection: customers consent by default to their data being used for marketing purposes. There is the possibility of an opt-out, which can be requested online after entering the card number. The business model with the data is constantly being refined. For example, issuers enter into partnerships with other companies ('co-branding' of cards) whereby the credit card can be used at the same time to collect air miles or bonus points when shopping online. A bond is created with the partner company and the customer's purchasing behaviour becomes traceable for both the issuer and the partner company.

The spread of credit cards is enormous—in the USA there are over 2400 credit cards per 1000 inhabitants (cf. ibid.). It is difficult to escape the use of plastic money and data collection. Without a credit card, one is left out of car rentals or online trading, for example. Correspondingly, companies have access to the consumption habits of practically the entire population.

The volumes of data from credit card payments are digitally exploited and turned into consumer tracking. Secrecy disappears from the payment process. A well-known result of this digital transparency is the automatic blocking of the card when the algorithm detects a suspicious payment: unusual location, amount, item. The assumption is then that the card has been stolen. For this security feature, a user profile exists that combines habitual locations and purchase items, times and amounts. If the card is not stolen, but the holder is just unaccustomed to travelling, he can call the hotline and try to explain the deviation from his own behavior. The knowledge that the card algorithm has gathered about the customer must be corrected in a personal conversation. The fact that the card is rarely blocked even for travel and rather unusual payments shows how precise the stored image is.

And the customer's map image in the cloud is not only needed for security purposes. With its help, spending profiles are created that are fed back to the customer. A pie chart shows which amount was spent in which consumption area per period and invites the optimization of the buying behavior. The tip of the data iceberg is made visible to the consumer. In addition, the data is actively used for tailored advertising offers and the optimization of the credit card model, even the personalized design of prices becomes possible through the knowledge of the willingness to pay (Seele et al. 2019).

The credit card has led to a sharp increase in the number of people involved in payments. The collected knowledge is efficiently managed and evaluated digitally. The number of users is constantly increasing thanks to innovative and low-threshold offers. The digital networking of payment is advancing: instead of cards, mobile phones and facial recognition can also be used for payment. In Sweden, these systems are so widespread that the Scandinavians aim to become the first cashless country in the medium term (see Zulauf 2018). The transition from cash to plastic money reduces privacy when making payments. The back of the cloud knows all receipts and knows about the next purchase before entering the store.

3.11.3 Cryptocurrency

With cryptocurrencies, anonymous payment could have taken a step into the digital world. But things turned out differently. It is true that digital currencies like bitcoin are not issued by a central authority like banks or central banks. Cryptocurrencies are decentralized, they are based on the blockchain. Simplistically, blockchains are databases attached to the currency that record every transaction made with the currency, similar to a collection of receipts. This database is stored on many computers and automatically updates with each use of the currency (Dierksmeier and Seele 2018, 2019). Through this public database, the transaction is authenticated: It confirms which uniquely identifiable participant handed out how much of their currency to whom. This is exactly where the difference with cash lies with regards to secrecy: The information about the payment transaction remains, becomes part of the ever-growing blockchain memory of the currency. The cryptocurrency has the digital co-knowledge firmly built in.

In addition to the technical side of transaction management, this co-knowledge fulfils a central task for the functioning of the currency: Bitcoins and other digital currencies make use of digital omniscience to establish trust. Instead of a central bank or a religious reference (think of the inscriptions on notes and coins from *In God we Trust* on the dollar to *Dominus Providebit* on the franc), digital omniscience guarantees that the tiny and abstract number on the screen is actually linked to a value and can be exchanged for goods. By having all transactions forever stored and retrievable in a decentralized manner on the blockchain, i.e. indelible and omnipresent, users can be confident that they will not be cheated out of their money. The cryptocurrency receives its metaphysical cover through the omniscience and omnipresence of the digital infrastructure (cf. Zapf 2018).

Of course, with cryptocurrency, you don't have to give your real name. Above all, it is anonymous addresses (wallets) to which the transactions are attributed and which are recorded in the blockchain. Due to the sheer mass of data and the pseudonymisation, the payment cannot be attributed to a real person for the time being (cf. Seele 2018a, b). However, this anonymity is lifted when cryptocurrency and the traditional monetary system intersect. Especially when converting money into digital currency. In order to comply with money laundering laws, depending on the national jurisdiction, exchange agencies have to collect personal data of currency customers above a certain amount—e.g. the name or a phone number. Thus, the secret is out and the link between virtual wallet and analog identity is established. This link to the analog world is currently spotty, as cryptocurrencies are largely unregulated. Once government regulation kicks in and the blockchain is no longer decentralized, the anonymous blockchain becomes a surveillance tool. Regulators will then have access to the cumulative knowledge of the currency and any other information that digital omniscience has accumulated in the blockchain (cf. Seele 2018a, b).

Control over the blockchain is not only of interest in matters of government regulation, but also for companies, which can use it to expand their access to the payment habits of their customers. Accordingly, companies are trying to issue their own cryptocurrencies to their customers—the best-known example is Facebook's currency called Libra, although it is not yet clear whether it will go beyond the trial phase.

Let's think back to the cash payment in comparison. Here, the data trail is complete when the note has made its way to the cash register. Digital payment is completely different. Whether with credit card or cryptocurrency: Here, the data processing only begins at the moment the money is handed over. Companies and governments collect the digital traces of the payment, evaluate them and thus take the secret out of paying.

3.12 Books and e-Books

The topic of 'books' shows in a vivid way what serious changes digitalization has on product design as well—and how little the customer himself needs to know about it. The knowledge of actualities, as applied on the back of the cloud, can also become the decisive criterion for which products are presented to a customer in an updated form. We feed the background information of this chapter from the genesis of this very book and its second edition.

3.12.1 One Edition, One Word

In past decades and centuries, books were written and eventually typeset and printed at considerable expense. Since the typesetting of a text was a profession in its own right—that of the typesetter—and each word had to be composed of individual letters at the beginning of book printing, the publisher gave the printer an approximate estimate of the number of copies to be sold and had an edition of a certain number of copies printed accordingly. Before the printing press, the only way of reproducing books was by hand, which was one of the main activities in monasteries in the Middle Ages. Then, with printing, reproducibility finally changed, and the cost of each additional copy that did not have to be copied individually dropped and dropped and dropped. This economy of scale of mass production was also an economy of scale for education. Books were no longer an esoteric luxury good, but could be produced cheaply and in large numbers. Nevertheless, the cost of a copy and a print run was a high entrepreneurial risk because human labor, production, machinery, materials, paper, and ink had to be used in a cost-conscious manner. If a book was a surprising success, a second edition could be arranged. Either one still had the text 'set', so one just had the printing presses produce it. Or—if the demand for a further edition came as a surprise—the typesetting, which was tidied up again after printing into the typesetting box of individual letters and punctuation marks, was set again.

Books were a luxury product in those days, despite the economies of scale. And since the saying goes: *What costs nothing is worth nothing!*, the publisher or editor had to make a thoroughly risky decision as to which manuscript (literally handwritten) was to become a highly reproducible book.

This museum-like form of book publishing, however, has been passé for decades, since the advent of electronic word processing for authors and electronic typesetting and printing presses. As a result, the basic cost of a book has dropped considerably and one can produce a higher number of books for less effort. The printing press, however, has to be set up, so one still has to make a sales calculation to estimate the likely number of books to be sold in X amount of time. However, a new edition is no longer dependent on the previously high production and set-up costs of the printing set.

3.12.2 Zeros and Ones Are Patient: Books and e-Books in Peaceful Co-Existence

With the proliferation of personal computers and eventually tablets and 'paper-ink' readers, the next logical step was consequently for publishers to add purely electronic books to the analogue ordeal of physical book production. With people spending more and more time in front of screens, it was only logical to also wrest the book from the paper medium and have the digital file accessed and read on screen. So 'e-book' formats were technically established whose purpose was not only to optimise the reading experience, but whose essential purpose was also to provide copy protection. If you look around in public, especially on public transport, in airports or at the beach, more and more people are using the electronic version of the book. This is produced once and can then be reproduced an infinite number of times at minimal cost. Walter Benjamin (1935) had described this principle long before the advent of the digital age in his famous book: "The Work of Art in the Age of its Technical Reproducibility". In the digital age, the marginal costs of technical reproducibility have finally reached almost zero. Let's just take the example of Walter Benjamin himself: his book is now freely accessible as an online version due to the expiration of copyright on the servers of Wikisource (Benjamin 1935). For the bibliophile reader, the work is also still available as a printed copy or as a paid-for, e-book-reader compatible format. The consumer decides, and as nice as this is for the consumer, and as good as this accessibility is for the theoretical possibility of self-education and further education, these technical possibilities are a great challenge for the publishers.

It is no wonder that it is the big publishers in particular who then sell fewer individual e-books (this too, of course), but rather entire subscription packages. The same applies to online retailers, where you can get access to electronic books for a monthly fee. The publisher of this book, for example, is in a position as both a general interest and academic publisher to offer access to e-books as a flat rate service. This is sold to university libraries or tied to memberships, which puts the reader in the comfortable position of downloading an entire book and deciding as they read whether to flip through the work, i.e. scroll, or save it to the patient expanses of the cloud. Printed book and online version thus live peacefully side by side. The bibliophiles and the old-fashioned can continue to buy the paper copy from their trusted bookstore or from an online bookseller. Those who have access to the package of books available online can download it to their device and read it—or have it read aloud—at no further cost. For the completely undecided, there are still reading samples that are provided digitally as excerpts.

3.12.3 E-Books: When the Reader Reads the Reader

However, the latest printing press technology is currently offering even more sophisticated possibilities. There is a new printer technology that makes it possible to print individual copies on demand. In addition to the electronic version, which is available as a purchasable e-book, as well as a package service for a framework agreement with the publisher where the customer is entitled to an inclusive download, a single book can thus be printed and sent on order. A novelty that can make the old logic of editions obsolete. But in the end, as the step of the co-existence of book and e-book has already shown, the edition is about marketing. Whereas in the analogue world it was the theoretical determination of demand that determined the size of the print run, nowadays it is the likelihood that the topic will generate new demand through a relaunch, through renewed publication. Particularly in the case of rapidly changing subject matter, another edition makes perfect sense simply because of the changing references, even if it is not a technically determined edition. In the case of the topic of digitality and privacy, it was the Cambridge Analytica case in particular, which we included in the first chapters, that raised the topic of privacy to a level of sensitivity that did not exist before. It also didn't exist because the technology and the will to manipulate most private beliefs, as seen in democratic voting, didn't exist. So a further edition today is no longer given by the limited nature of the first edition, but rather an instrument for books whose subject matter changes quickly and whose topicality makes the prospect of further sales success seem likely. With digital books, the limitation is not there anyway, since zeros and ones are patient as long as they get some energy. So essentially it remains a matter for publisher marketing to decide when it is time for a second edition. This is best done when the marginal cost of producing the physical book equals the unit cost. But even this can be overcome by the next trick of digital bookmaking: "living editions" of books allow for continuous updating without the need for an explicit second edition. But for reasons of topicality and marketing, the new edition still has a special significance, and not only for reasons of updating.

There are also other reasons for the future of the book to be digitized. Reasons that bring us closer to the actual topic of the back of the cloud. With an e-book read on a digital e-book reader, not only does the reader read the book, but the device also reads the reader. Not only is the book related to the reader in its genre-specific genre. Reading speed and pauses at particular parts of the text are also read out by the e-book-reader and, where possible, related to the data of the digital self. So if the book is traditionally used for personal development and education, because it challenges the reader cognitively, the digital variant adds a completely new facet. Who reads which books and when, and which passages are read in detail and which in passing, allows conclusions to be drawn about the person and their character traits. This can also be used, for example, in the interests of the security services. People who read books about terrorist attacks and load and read books about chemicals or bomb-making can thus be quickly identified as threats by algorithmic intelligence. But it could all be quite different. Someone is studying chemistry and reading technical books for their studies. At the same time, the person is in a reading circle where a terror thriller is being read. The choice of reading, however, has nothing to do with the specific choice of the person reading, but solely with the contingent fact of being a member of a reading circle.

Stored on the back of the cloud is thus the reading and evening behavior. How long does someone read until they fall asleep in the evening? Which books and which chapters and actions keep people busy and which ones—aggregated over millions of data records—make them put the e-book down. Which product placement does well when correlated somewhat with the person's consumption behavior on the Internet or via GPS data from the smartphone in a shopping street.

While the book used to be an exclusive instrument of the freedom of the mind, of the formation of thoughts, as dystopically thematized in Ray Bradbury's Fahrenheit 451, the perspective is now mirrored and reversed. Thoughts are free—once upon a time. But the forming of thoughts is already being captured and processed on the back of the cloud as well.

3.13 Sexuality and the Internet: The Incognito Illusion

This chapter is arguably the most important chapter in this new edition, because the technologies it introduces for mining private and most private content and activities are now so perfectly perfidious: To make this point as vividly as possible, we use one of marketing's most common principles: sex sells. Arguably, there is no place where the structural transformation of privacy can be better demonstrated: both in terms of the surveillance of actual sexuality in private space. Here we are talking about smartphones that are bugging devices, TV sets that eavesdrop, fitness tracker bracelets that can automatically detect and classify physical activity. And we are talking, secondly, about the monitoring capabilities of the consumption of the representation of sexual acts on the Internet: Subcontractors providing data that the big audience companies won't/aren't allowed to provide. The ambiguous protection of the virus scanner that constantly reads everything (it has to watch all the time, after all) and eventually resells the sensitive data. Networked TV devices that can eavesdrop (and in some cases watch) what is being played in the room, even after the main switch has been pressed ("fake-off mode"). Incognito modes of browsers, which smart subprograms look over their shoulder, so they are not incognito after all.

And to make sure that the knowledge about the infamous tricks spreads especially well, the practical example of this chapter is dedicated to the exploration of human sexuality by the spies on the back of the Cloud, who are inspired by fantasies of omnipotence.

3.13.1 Adult Entertainment from the Station Bookshop

There has always been a lot of running at the main station. Thousands of people hurry on their way. In the streams of the crowd, moreover there are many travelers from out of town, mostly in the anonymous centre of inner cities, it was previously possible to buy a product unrecognised, the act of purchase of which not everyone in their own city wants to be seen doing. What is technically called "adult entertainment" in the financial sector (Jensen and Seele 2013) includes products from the areas of: pornography, tobacco products or alcoholic beverages. Products that are beholden to a legal minimum age restriction. The proverbial smut already contains a pejorative connotation of social control and social undesirability, but this does not seem to have done much damage to demand. Similar to the previous examples of the analogue world, a shopper walks into the store with no known identity, possibly even still wearing a turned-up collar and cap, chooses a product, pays for it with cash, and walks away. The transaction remains fully private—even secret. Inclinations and preferences remain secret and unrecognized. The only pitfall of exposure could still have been the age check. The effort, on the other hand, of joining the anonymous crowd, going into the shop and making a purchase are clearly an expression of the old analogue world. Anyone who has eaten in Frankfurt am Main at the excellent South Indian restaurant near the train station on the legendary Kaiserstraße will know that the shop in the building next door deals in products clearly that are recognisable as "adult entertainment", but operates under the brand and company name "Dr. Müller". Social acceptance on the receipt guaranteed.

3.13.2 When Pictures Learned to Surf

With the introduction of electronic payment by credit card, however, the degree of secrecy of the transaction changed. "Dr. Müller" obviously successfully anticipated this with this already doubly unsuspicious name, which should appear on every credit card statement. The more significant change from analog to digital, however, was the change in the medium of presentation from print to digital images. The proverbial dirty booklet is in print and as we know: Paper is patient. And blind, deaf and dumb. With the media shift to the world of digital pixels, image and sound content became playable or retrievable on any device. While in the early days of digitality a computer was still primarily a pure playback device, in the last decades the technologies developed more and more in the direction that the playback device began to *know* who was playing what. If content was accessed on the Internet and paid for by credit card, for example, data was also available to the card companies. The question of social acceptance, which will have changed socially just as much as the technology, thus shifted from one's own social group of the street or city to the privacy and physical enclosure of one's own four walls—with the simultaneous digital opening of the physical enclosure through digital connectivity.

3.13.3 The Incognito Illusion, Fitness Trackers and Bedside Bugs

In order to develop a theory of privacy that works without the fundamental principle of secrecy in the digital age, the topic of sexuality is best suited to illustrate the changes of digital transformation.[5] The caesura for our understanding of privacy is so serious, in fact, that the Internet can be divided into the following three phases in order to better understand the loss of secret privacy:

Phase 1 consisted of ubiquity, the omnipresence of information that became available anytime, anywhere. The Internet was more of a one-way flow of information for the user, except for some hacker attacks. Emails, like any other communication of the prehistoric era, could be intercepted but were a

---
[5] The following deals exclusively with matters irrelevant to criminal law.

communication between individuals. Overall, the Internet was a large, interconnected library that no longer held books, but data.

Phase 2 began when the Internet of utilities like algorithms and cookies emerged, actively and relentlessly collecting data about users this side and that side of legalities and automatically analyzing it for pattern recognition. The great real-time lexicon of the Internet began to take an interest in the human ends of communication flows. Not 'The Internet' per se, of course, but the private sector organisations behind it and—as has always been the case—the services of governments. The new Data Protection Regulation has even helped for Phase 2, in that everyone on almost every site now has to actively allow Phase 2 secret agent algorithms first. You can't use many companies and sites at all otherwise. Legally, this lets you off the hook as a company, since the company has obtained a personal consent form to a set of contracts that almost no one even reads. But the greed for data, for truly personal data, is insatiable. Not only are users lured into a psychological addiction by autoplay continuities, by nudges of who from my social network just said what about what. The data troves of social media such as Facebook, Twitter, Whatsapp, Instagram, Amazon, and Alphabet were not enough (Graham et al. 2017). Perhaps not unfiltered enough even among some reluctant users. The so-called panopticon effect states that people behave differently under observation. You have scissors in your head, and say and do different things than when you feel safe and 'at home' in the privacy of your own free thoughts.

Therefore, one should still address a *phase 3:* The time when companies started to maintain the simulation of privacy, but at the same time used subversive and perfidious tricks on this side and on the other side of legality to indirectly collect data through bonds, through third parties or by making false assumptions like a simulated main switch ("fake-off switch"), when the user thought he was alone, he was by himself, he was private. Completely private, just with himself. Convinced that the time-honored understanding of privacy in conjunction with the secret was the realm in which one is undisturbed, intimate (Fig. 3.4).

In the following, therefore, this chapter presents the execution of this dramatic change of privacy on the subject of sexuality. On the one hand, through the technical devices that surround us as humans in our homes—and that are capable of taking away our secrecy and privacy. Secondly, through the technological devices when it comes to the consumption of sexual content online, a rather direct but from the data side probably even more 'interesting' facet, as digital companies can thus learn about inclinations and desires in different ways other than classic online shopping. So there is much for digital companies to discover in what is, according to American communications scholar

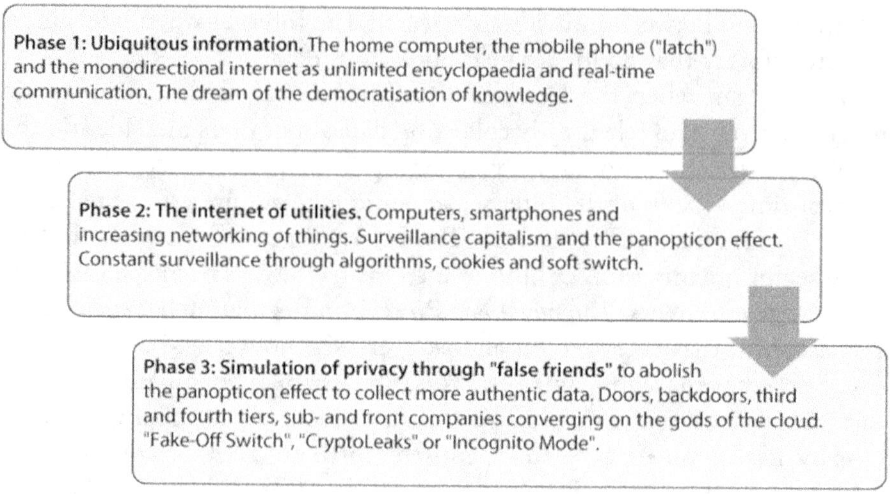

Phase 1: Ubiquitous information. The home computer, the mobile phone ("latch") and the monodirectional internet as unlimited encyclopaedia and real-time communication. The dream of the democratisation of knowledge.

Phase 2: The internet of utilities. Computers, smartphones and increasing networking of things. Surveillance capitalism and the panopticon effect. Constant surveillance through algorithms, cookies and soft switch.

Phase 3: Simulation of privacy through "false friends" to abolish the panopticon effect to collect more authentic data. Doors, backdoors, third and fourth tiers, sub- and front companies converging on the gods of the cloud. "Fake-Off Switch", "CryptoLeaks" or "Incognito Mode".

Fig. 3.4 The three phases of the Internet's erosion and distortion of privacy

David Slayden, "the quietest big business in the world", as Dietmar Dath points out in the F.A.Z. (Dath 2014). And there is all the more to discover, the more secure, private and unsupervised the user believes his behaviour to be. And we are talking about a big pie of data: One-third of the traffic on the internet in 2019 would be online pornography. This data traffic alone would also be responsible for CO_2 emissions equivalent to the annual value of a country like Romania (Holland 2019). But these are so-called externalities. What about the user and how private or "secret" is their media consumption? Here are some of the examples that show exactly how everyone is spied on today. Spying, unfortunately, is the correct word.

Online consumption: By now, word should have spread—also due to the EU's data protection regulation—that our viewing and usage habits are monitored and read out in great detail. Under surveillance, we do not behave freely, the so-called scissors in the head of the panopticon effect (Seele 2016). For example, a study by Maris et al. (2019) titled "Tracking Sex" found that in 22,484 investigated sites with pornographic content, 93% of them forwarded user data to third parties. And this was done in the "incognito mode" of the web browser. This "incognito mode", according to the study, only ensures that the search history is not stored on the computer. Nothing more. However, the mode that appears as 'incognito' still potentially collects activity on the computer and allows third parties and their utilities (cookies, algorithms) to profile the user's sexual preferences based on the web address and its meaningful titles. Maris et al. put it succinctly, "the reassurance of the 'incognito' mode icon on his screen, provide Jack [a fictional user of the study] with a fundamentally misleading sense of privacy as he consumes porn online" (Maris et al. 2019, p. 1). The term 'incognito' is thus misleading. At the very least, it must be stated.

In addition, the study authors analyzed the privacy policies and data protection guidelines of the companies and found that 44.97% associate a 'sexual identity' and 'preferences' with the user based on usage patterns. However, what might be even more disturbing is the finding from the study that the major internet companies that almost all of us deal with on a daily basis are using this data in their analytics, even though they don't necessarily collect it themselves in the first instance. For example, the authors found that Google or its subcontractors have trackers, colloquially small spyware, on 74% of all pornographic websites. Oracle still has 24%, and Facebook, which itself does not tolerate any nudity or pornography on its sites, still tracks a tenth of all pornographic sites. *(Nota bene: Microsoft is not listed in the study. Whether this is because the first author is listed in the publication as working at Microsoft remains to be seen).*

If one does not necessarily start from the existing remnants of a free, open society, but considers ideological regimes, dictatorships or other systems of oppression, it becomes clear that these data are highly explosive. Dirk Helbing (2020) has shared an apt observation on privacy: "A society without privacy will either end as totalitarian or a shameless society. Privacy is particularly important, because minorities are always vulnerable, but society depends on many minorities. Entrepreneurs, politicians, judges, intellectuals and artists are all minorities, and they are vulnerable". This point is the ultimate criterion for privacy and its protection. In history, we have seen in too many places the persecution of people who were either persecuted, imprisoned or even killed for sexual, political or religious beliefs. That is what we are dealing with. But let us turn to the other alternative of Helbing's dimensions of abolished privacy: the shameless society.

Now one can say: the Internet companies only want to earn money and have nothing to do with repression and persecution or totalitarianism, because it is 'only' about profits. And if you sell advertising, then a better, more intimate knowledge is definitely an advantage. It is well known in marketing, for example, that homosexuals are statistically more likely to work in creative professions and statistically more likely to have higher net incomes. So if you had that information, you could advertise in a much more targeted way and also assume a higher willingness to pay on the part of the consumer. So for commercial value creation, sexual orientation is definitely relevant. So let's look at the technical possibilities of misleading users to obtain their private, secret, intimate data, which are characteristic of phase 3. The range of possibilities and the tools of darkness in the refinement by tricks and gimmicks is immense. And that's understandable: for one thing, customer profiles are a billion-dollar market. For another, laws, even the new EU Data Protection Regulation, lag behind the possibilities of reality. On a related note, Maris et al. also found that in 17% of cases, privacy policy documents are written in language that requires a two-year college degree to understand. Officially, Google and Facebook commented on the study that they do not collect user data on sensitive content themselves. They don't have to, as various cases of so-called add-ons show. The most inglorious case was probably the Nacho Analytics case, where an analytical add-on sold confidential data of millions of users on the Internet. The core of the scandal is the ignorance of users that their data, including highly private information such as tax records and medical data, is being sold as a commodity (Langer 2019).

Now you would think that the situation would have improved with the European Data Protection Act. But by the fact that companies now have to obtain a declaration of consent from the users—and the users, if they want to use some sites, have given their consent, the companies have carte blanche. After all, who reads the smallest grey print on white contracts? Some sites even set an all-or-nothing rule. If you don't agree, you can't use the site. It's as simple as that. And in the process reveals the bargaining power and platform gravity of some big data companies. Tobias Gostomzyk (2020) calls this the "great consent lie" or a "decision fiction." The staircase joke of the sovereign data citizen is that he gives his consent in this "big consent lie" and is therefore legally allowed to be monitored and read on the back of the cloud. Chapeau! The best trick, however, has been devised by the anti-virus provider AVAST: in order to be protected against viruses, it is in the nature of things that every threat, no matter how small and hidden, must be detected. In other words, the idea of anti-virus protection is that it must literally shine a light into every last nook and cranny of both the computer, the files stored there, the programs, the emails and attachments, the sites visited and their utilities. AVAST grasped the enormous potential of this total screening and, according to a report in Motherboard (Cox 2020), created its own market for web browsing data. Their clients were big-name companies in various industries, including Google, Microsoft, Pepsi, and McKinsey. User data was advertised with the slogan: "Every search. Every click. Every buy. On every site." And this brings us full circle to the profiling of users and their most secret, private and intimate preferences. Collected and analyzed by the army of subservient algorithms at the back of the cloud, the virtual agents of the legal persona of these companies that ultimately spy on us, the natural persons, and the nature of our private needs. Cox reports hundreds of millions of users around the world. Again, the culprit was not the company's well-known flagship product, but a sub-unit called "Jumpshot." According to the report, the product that the big-name customers were able to purchase is called "All Clicks Feed," which is a sort of log of all clicks made on the computer. And yes. All Clicks literally means all clicks. According to Cox, this includes Google search queries, room coordinates on Google Maps, Linked-In profiles, YouTube videos, and the anonymized hits with time and location from sites like YouPorn or PornHub, as well as, in some cases, the explicit search terms and the specific videos, according to Cox. That's what we're dealing with.

The striking saying is true: We human users are no longer the customers of Internet companies. We a r e the product. And our most intimate preferences are a product property that adds value to the data set. We are modeling our digital double for the companies. But it gets better:

Analogue sexuality, digitally captured. Since, as mentioned at the beginning, about a third of the Internet data volume is of pornographic nature, one can say: well, okay. It affects a lot of people, but those digitally abstinent analogue advocates are not covered. But this is not the case. A range of devices that people have around them all the time, even in the most private of moments, captures live sexuality. If you were to delve into the depths of the operating system settings, if you were to read the usage agreement of apps, you would be able to see—some of this is listed on the summaries—that some core, indispensable apps demand access to the microphone as a condition of use. The same applies to motion sensors in the device that react to vibrations and movements—you might know this from the automatic image rotation function when you move the device from vertical to horizontal and the image jumps around.

Those who use pattern recognition algorithms well can get a pretty accurate picture of what's going on via facial and voice recognition. One notable case embarrassed Facebook in 2017. Matney (2017) reports a video that went viral of a young couple who apparently had doubts about whether Facebook was listening to them. So they conducted an experiment as simple as it was intelligent. With the phone on the table, they talked about cat food several times a day and they reminded each other to urgently buy cat food the next day. Promptly, the device showed several ads for cat food the following day. The punchline is: They both did not and do not have a cat. They vividly filmed this with a camera, and journalist Matney lets Facebook respond: "Facebook does not use your phone's microphone to inform ads or to change what you see in News Feed," Facebook wrote in a blog post last year. "Some recent articles have suggested that we must be listening to people's conversations in order to show them relevant ads. This is not true. We show ads based on people's interests and other profile information—not what you're talking out loud about." (Matney 2017). So it's testimony against testimony and that usually means: someone is lying. Or no one is. If you take the company's statement and relate it to what it says above about third party add-ons and utilities reporting back to the big internet companies, then the statement is formally correct (the interersssive and cloistered part of the sentence is: "and other profile information"—which can mean many things, possibly including third party utilities). No need to lie at all, a strategically placed ambiguity or vagueness, backed by a web of opaque but legal and consent-approved services, solves the embarrassment to the satisfaction of product developers and shareholders.

Particularly inquisitive little computers are, for example, fitness wristbands or watches. Here, too, a heart rate monitor, pedometer, vibration sensor and more have been built in to promote the user's health. Anyone who has ever worn such a device might know that the algorithm stored in the cloud, to which the fitness tracker is connected via app on the smart mobile phone via the internet, can read all sorts of things from it. Good devices can easily tell what sports activity the wearer is doing. Walking, hiking, jogging, swimming, cycling or other. If you look in the menu control of the activities to be selected, you will find a range of over a dozen sports with some manufacturers. It is hard to imagine that the specific sensor data of sexual activity cannot be read and recognized by the device. Probably, it is thanks to the reverence or just business acumen of the companies that users are spared this category. But on the back of the cloud, might it not still be collected, recognized, sent to the app, sent to the cloud, and stored in the user profile of the digital double in the cloud? And may we rule out the possibility that there are no third party add-ons that intelligently evaluate the products and link back? And what about GPS or near-field tracking? Can't you possibly identify your partner and their movement and biometric activity profile as well?

One might conclude that the absence of all smart devices could provide a residue of privacy that would even be secret. But since almost everything is smartified today, and it is more likely that the trend will continue. Perhaps an evening of television, or even just an evening of reading in the living room, is in front of our subjects' eyes. Computers, phones, fitness trackers stay in the next room with the door closed. Let's imagine there's just a sofa, a table and a TV. Maybe one from Samsung, they're supposed to have high picture quality at a reasonable price. But here, too, the algorithm trap snaps shut. De Leuw reported (2017) on snooping Samsung TV sets. Thanks to the disclosure platform Wiki-Leaks, we know about a hack called "Weaping Angel", which was installed externally on the TV and turns on the built-in microphone during a so-called "fake-off mode", i.e. when the device is supposedly switched off, it listens in and sends data via Wi-Fi.

Any way you slice it: Legal, illegal, voluntary, involuntary. We are being monitored and our most secret, private, intimate moments are—unless one is inclined to practice extreme digital asceticism or digital detox (and has the possibility to do so at all)—no longer private. And the secret has become a variable in a data set to better market us, the product for the digital giants.

So, based on these detailed case histories, it is now time to theoretically recast the notion of privacy to reflect the serious changes brought about by the algorithms and gods on the back of the cloud, and to have an updated and more real concept of privacy.[6]

References

Balz, Dan (2013): How the Obama campaign won the race for voter data. *The Washington Post*, Politics, July 28.
Barrett, David (2016): Google to deliver wrong ‚top' search results to would-be jihadis. *The Telegraph*, Technology, February 2.
Baum, Karl-Heinz (2002): Wenn schon vorher alles klar ist: Auch Diktaturen lassen gerne wählen. *fluter*. http://www.fluter.de/de/wahlen_special/editor/839/
Beeger, Britta (2014): 7 Dinge, die Sie über facebook nicht wissen. *FAZ*, Wirtschaft, February 3. http://www.faz.net/aktuell/wirtschaft/netzwirtschaft/der-facebook-boersengang/zehn-jahre-facebook-7-dinge-die-sie-ueber-facebook-nicht-wissen-12782981.html
Benjamin, Walter (1935). *Das Kunstwerk im Zeitalter seiner technischen Reproduzierbarkeit.* https://de.wikisource.org/wiki/Das_Kunstwerk_im_Zeitalter_seiner_technischen_Reproduzierbarkeit_(Dritte_Fassung) (aufgerufen am 27. Januar 2020).
Benney, Jonathan (2013): The aesthetics of microblogging: How the Chinese state controls Weibo. *Tilburg Papers in Culture Studies* (66), S. 0–20.
Benoist, Jean-Christophe (2012): *Time Square – From upperstairs.* Copyright-frei nach CC-BY-SA-3.0. https://commons.wikimedia.org/wiki/File:NYC_-_Time_Square_-_From_upperstairs.jpg. Auth. Benoist, Jean-Christophe. (Hrsg.) (2012): Wikimedia Commons.

---

[6] Who is also interested in what practical possibilities there are to cover the digital tracks or even distort them. The first author of the book collects loose recommendations and tips for civil disobedience in the digital. To that end, there's a Twitter account inspired by Gandhi's passive resistance called: "Digital Salt March" @digital_march.

Bernstein, Joseph (2015): Ashley Madison's $19 ‚Full Delete' Option Made The Company Millions. *Buzzfeed.* Abrufbar unter: http://www.buzzfeed.com/joseph-bernstein/leaked-documents-suggest-ashley-madison-made-millions-promis?bftwnews&utm_term=4ldqpgc#.naXL9ddqj *(letzter Zugriff: 20.08.2015).*

BILD (2015): Mediadaten – BILD Zeitung. Abrufbar unter: http://www.media-impact.de/dl/18562799/BILD_PL_2015_04.11.14.pdf *(letzter Zugriff: 10.11.2015).*

Bose, Anirban and Bruno Mellado (2019): *World Payments Report 2018.* Paris: BNP Paribas/Capgemini.

Botsman, Rachel; Rogers, Roo (2011): *Whats mine is yours. The Rise of Collaborative Consumption.* New York: Harper Collins.

Bridle, James (2014): The algorithm method: how internet dating became everyone's route to a perfect love match. *The Guardian,* Valentine's Day, February 9.

Bruckmann, Claudia and Martina Eschelbach (2018): *Zahlungsverhalten in Deutschland 2017.* Frankfurt a.M.: Deutsche Bundesbank.

btl creative (2015): Mission. *btl creative – below the line.* Abrufbar unter: https://www.btl-creative.com/#section-mission *(letzter Zugriff: 08.12.2015).*

Büchel, Jeremias (2015): „Bringt euer Essen selber mit oder bleibt hungrig". *20 Minuten,* Generation Geiz, July 23.

Bütler, Monika (2015): Abstimmen per Knopfdruck verändert die Entscheide. *NZZ am Sonntag,* Schweiz, October 4.

civity (2014): Urbane Mobilität im Umbruch. *Matters No. 1.* Abrufbar unter: http://matters.civity.de *(letzter Zugriff: 18.01.2016).*

Cox, Joseph (2020): Leaked Documents Expose the Secretive Market for Your Web Browsing Data. https://www.vice.com/en_us/article/qjdkq7/avast-antivirus-sells-user-browsing-data-investigation (200212).

Crueger, Jens (2013): Privatheit und Öffentlichkeit im digitalen Raum: Konflikt um die Reichweite sozialer Normen. *Aus Politik und Zeitgeschichte* 15–16 S. 20–24.

Dallmer, Hans (2015): Lemma: Direktmarketing. *Gabler Wirtschaftslexikon.* Abrufbar unter: http://wirtschaftslexikon.gabler.de/Archiv/618/direct-marketing-v10.html *(letzter Zugriff: 11.11.2015).*

Dath, Dietmar (2014): Im Weltreich der nakten Daten. *FAZ.* https://www.faz.net/aktuell/feuilleton/medien/internetpornographie-im-weltreich-der-nackten-daten-12764666.html (200115).

Dierksmeier, Claus; Seele, Peter (2018): Cryptocurrencies and Business Ethics. Journal of Business Ethics 152.1 1–14. https://doi.org/10.1007/s10551-016-3298-0 (selected by Springer Editors-in-Chief in the collection: „Change the World").

Dierksmeier, Claus; Seele, Peter (2019): Blockchain and Business Ethics. Business Ethics: European Review. https://doi.org/10.1111/beer.12259

Dribbusch, Barbara (2013): Die Neidspirale. *taz,* Kultur, January 21.

Dworschak, Manfred (2015): Supereinfach! Superlecker!. *DER SPIEGEL* (50), S. 110–113.

Eilert, Bernd (2011): Disney-Stadt Celebration: Kein Paradies ohne Schlange. *FAZ*, Feuilleton, February 16.

Fabricius, Michael (2015): Wenn Mieter sich im Internet selbst versteigern. *DIE WELT*, Wohnungssuche, March 25.

facebook (2016): Was sind facebook custom audiences? Abrufbar unter: https://de-de.facebook.com/business/help/341425252616329 *(letzter Zugriff: 12.05.2016)*.

Falkvinge, Rick (2015): In China, Your Credit Score Is Now Affected By Your Political Opinions – And Your Friends' Political Opinions. *privateinternetaccess blog*. Abrufbar unter: https://www.privateinternetaccess.com/blog/2015/10/in-china-your-credit-score-is-now-affected-by-your-political-opinions-and-your-friends-political-opinions/ *(letzter Zugriff: 06.10.2015)*.

Fox-Brewster, Thomas (2015): Location, Sensors, Voice, Photos?! Spotify Just Got Real Creepy With The Data It Collects On You. *Forbes Magazine*. Abrufbar unter: http://onforb.es/1KxiVX6 *(letzter Zugriff: 20.05.2015)*.

Franzen, Jonathan (2015): *Unschuld*. Reinbek bei Hamburg: Rowohlt.

Graham, M., Hjorth, I., & Lehdonvirta, V. (2017). Digital labour and development: impacts of global digital labour platforms and the gig economy on worker livelihoods. *Transfer: European Review of Labour and Research*, 23(2), 135–162. https://doi.org/10.1177/1024258916687250.

Garrett, Sean (2015): Early Twitter – Early Meerkat. *Twitter.com*. Abrufbar unter: https://twitter.com/SG/status/574755572542652416 *(letzter Zugriff: 16.07.2015)*.

Goode, Erica (2011): Sending the Police Before There's a Crime. *The New York Times*, U.S., August 15.

Gostomzyk, Tobias (2020): Die große Einwilligungs-Lüge. https://www.sueddeutsche.de/digital/tracking-einwilligung-datenschutz-meinung-cookies-1.4760678 (200212).

Gotthelf, Jeremias (1838): *Leiden und Freuden eines Schulmeisters*. Bern: Julius Springer.

Gröschner, Rolf (2013): Wählen gehen – öffentliche Angelegenheit des ganzen Volkes. *FAZ*, Politik, July 25.

Grundlehner, Werner (2018): „Im Schliessfach feiert das Bankgeheimnis Urständ", NZZ, Finanzen, 21.3.

Gutknecht, Mario (2015): Kanton Aargau: Dorfpolizisten sterben aus. *SRF*. Abrufbar unter: Kanton Aargau: Dorfpolizisten sterben aus *(letzter Zugriff: 24.11.2015)*.

Hagelüken, Alexander; Jalsovec, Andreas (2011): 15 Jahre Volksaktie – 15 Jahre Leiden. *Süddeutsche Zeitung*, Wirtschaft, November 18.

Hawksworth, John; Vaughan, Robert (2014): The sharing economy – sizing the revenue opportunity. *PWC.com*. Abrufbar unter: http://www.pwc.co.uk/issues/megatrends/collisions/sharingeconomy/the-sharing-economy-sizing-the-revenue-opportunity.html *(letzter Zugriff: 19.01.2016)*.

Helbing, Dirk (2020): Privacy. https://twitter.com/FuturICT/status/1217764449290985472 (200116).

Hellman, Peter (1997): Bright Lights, Big Money. *New York* 30 (19), S. 46–51.
Holland, Martin (2019): Klimawandel: Online-Pornos produzieren so viel CO2 wie Rumänien. https://www.heise.de/newsticker/meldung/Klimawandel-Online-Pornos-produzieren-so-viel-CO2-wie-Rumaenien-4469108.html (29.01.2020).
Hui, Julie S; Greenberg, Michael D; Gerber, Elizabeth M (2014): Understanding the role of community in crowdfunding work. *Proceedings of the 17th ACM Conference on Computer Supported Cooperative Work & Social Computing* S. 62–74.
Jahr Top Special Verlag (2015): Fliegenfischen – Mediadaten. *FliegenFischen*. Abrufbar unter: https://www.jahr-tsv.de/marken/fliegenfischen *(letzter Zugriff: 11.11.2015)*.
Jensen, Ole; Seele, Peter (2013): An Analysis of Sovereign Wealth and Pension Funds' Ethical Investment Guidelines and Their Commitment Thereto. In: Journal of Sustainable Finance and Investment.13/3/3 264–282. https://doi.org/10.1080/20430795.2013.791144
Kennedy, Pagan (2013): Who Made Speed Dating? *The New York Times Magazine*, September 20.
Knibbs, Kate (2015): Woman Says She Got Fired for Deleting a 24/7 Tracking App Off Her Phone. *Gizmodo*. Abrufbar unter: http://gizmodo.com/woman-says-she-got-fired-for-deleting-a-24-7-tracking-a-1703757185 *(letzter Zugriff: 12.05.2015)*.
Knupfer, Gabriel (2019): „Am Freitag endet die Ausgabe der 500-Euro-Note", Handelszeitung, Geld, 25.4.
Langer, Marie-Astrid (2019): Ausspioniert vom eigenen Internet-Browser. https://www.nzz.ch/panorama/ausspioniert-vom-eigenen-internet-browser-ld.1497200 (120212).
Leuthold, Karin (2015): Mann sonnt sich auf Windturbine. *20 Minuten*. Abrufbar unter: http://www.20min.ch/panorama/news/story/Mann-sonnt-sich-auf-Windturbine-17711979 *(letzter Zugriff: 31.08.2015)*.
Lobe, Adrian (2015): Im Netz der Wahlkampfhelfer. *FAZ*, Feuilleton, August 6. http://www.faz.net/aktuell/feuilleton/debatten/die-digital-debatte/algorithmen-beeinflussen-politische-willensbildung-13735791.html#void
Lovink, Joseph (2011): Die Gesellschaft der Suche: Fragen oder Googeln. *bpb*. Abrufbar unter: http://www.bpb.de/gesellschaft/medien/politik-des-suchens/75882/fragen-oder-googeln *(letzter Zugriff: 10.02.2016)*.
Lyon, D. (2014). ‚Surveillance, Snowden, and Big Data: Capacities, consequences, critique'. *Big Data & Society*, 1, 1–13.
Maass, Dave (2016): „No Cost" License Plate Readers Are Turning Texas Police into Mobile Debt Collectors and Data Miners. *Electronic Frontier Foundation*. Abrufbar unter: https://www.eff.org/deeplinks/2016/01/no-cost-license-plate-readers-are-turning-texas-police-mobile-debt-collectors-and *(letzter Zugriff: 27.01.2016)*.
MacCannell, Dean (2011): *The ethics of sightseeing*. Berkeley: University of California Press.

Maris, E.; Libert, T, und Henrichsen, J. (2019): Tracking sex: The implications of widespread sexual data leakage and tracking on porn websites, https://arxiv.org/abs/1907.06520

Matney, Mandy (2017): Facebook says it's not listening to you through your mic – but some users don't buy it. https://www.miamiherald.com/news/nation-world/national/article181947111.html (200212).

Matussek, Matthias; Oehmke, Philipp (2007): Die Tage der Kommune. *DER SPIEGEL* (5), S. 136–152.

Meskens, Ad (2010): *Martin of Tours at Basel Munster.* Copyright-frei nach CC-BY-SA-3.0. https://commons.wikimedia.org/wiki/File:Martin_of_Tours_at_Basel_Munster.JPG. Auth. Meskens, Ad. (Hrsg.) (2010): Wikimedia Commons.

Meusers, Richard (2015): Gestatten, RoboKoch. *Spiegel Online.* Abrufbar unter: http://www.spiegel.de/netzwelt/web/kochroboter-moley-robotics-lehrt-maschine-kochen-a-1066971.html *(letzter Zugriff: 16.12.2015).*

Mohrmann, Ruth-E. (2012): Zur Geschichte des Schlafes in volkskundlich-ethnologischer Sicht. *Rheinisch-westfälische Zeitschrift für Volkskunde* 57 S. 15–34.

NDR (2015): Schöne neue Welt – der Preis des Teilens. *Transkript zur Sendung Panorama, 08.01.2015.* Abrufbar unter: http://daserste.ndr.de/panorama/archiv/2015/panorama5378.pdf *(letzter Zugriff: 17.09.2015).*

Nelson, Shellie (2016): Debate goes federal over so-called ‚spying billboards' that track people who pass by them. *WQAD.* Abrufbar unter: http://wqad.com/2016/05/01/debate-goes-federal-over-so-called-spying-billboards-that-track-people-who-pass-by-them/ *(letzter Zugriff: 09.05.2016).*

Newman, Mark (2010): *Networks: An Introduction.* Oxford: Oxford University Press.

Owyang, Jeremiah (2014): Collaborative Economy Honeycomb 2 – Watch it Grow. *WebStrategist.* Abrufbar unter: http://www.web-strategist.com/blog/2014/12/07/collaborative-economy-honeycomb-2-watch-it-grow/ *(letzter Zugriff: 18.01.2016).*

padeluun (2014): BigBrotherAward 2014 für LG Electronics. *BigBrotherAwards.de.* Abrufbar unter: https://bigbrotherawards.de/2014/verbraucherschutz-lg *(letzter Zugriff: 03.08.2015).*

Pariser, Eli (2011): *The filter bubble: What the Internet is hiding from you.* New York: Penguin.

Partyguerilla (2013): AGB. *partyguerilla Deutschland.* Abrufbar unter: http://partyguerilla.com/de/unternehmen/agb/ *(letzter Zugriff: 25.03.2015).*

Person, Daniel (2015): Amazon Is So Nice to Employees, It Makes Your Personnel Information Public If You Criticize It. *Seattle Weekly.* Abrufbar unter: http://www.seattleweekly.com/home/961299-129/amazon-is-so-nice-to-employees *(letzter Zugriff: 20.10.2015).*

Pitzke, Marc (1999): „Als Frank McNamara seine Brieftasche vergass", NZZFolio, Das liebe Geld, Ausgabe 07.

Pollak, Sabine (2013): *Wohnen und Privatheit. Materialien zur Vorlesung im Modul Wohnbau 13/14.* Wien: TU Wien.

publisuisse (2005): Fernsehwerbung Wirkt Auf Jeden Fall, Aber Wie Und Warum Wirkt Der Einzelne Fernsehspot?. *Impact Dossier.*

Pulliam-Moore, Charles (2015): This malicious Android app lures you in with porn, takes your picture, then shakes you down for money. *fusion.* Abrufbar unter: http://fusion.net/story/194408/adult-player-android-ransomware/ (letzter Zugriff: 20.10.2015).

Reimer, Jule (2015): Gerichtsurteil Hamburg Nicht alle Partnerbörsen dürfen Geld nehmen. *Deutschlandfunk.* Abrufbar unter: http://www.deutschlandfunk.de/gerichtsurteil-hamburg-nicht-alle-partnerboersen-duerfen.697.de.html?dram:article_id=317927

Rogers, Adam (2015): Google's Search Algorithm Could Steal the Presidency. *Wired Magazine,* Science, June 8. http://www.wired.com/2015/08/googles-search-algorithm-steal-presidency/

Schlesiger, Christian (2014): BMW hat bei Carsharing die Nase vorne. *WirtschaftsWoche,* Auto, August 18.

Schäffler, Frank (2019): „Giralgeld, Bargeld, Kryptogeld", Frankfurter Allgemeine, Finanzen, 26.07.

Seele, Peter (2016): Envisioning the digital sustainability panopticon: a thought experiment of how big data may help advancing sustainability in the digital age. *Sustainability Science.* 11(5), 845–854 https://doi.org/10.1007/s11625-016-0381-5

Seele, Peter (2017). Predictive Sustainability Control: A review assessing the potential to transfer big data driven ‚predictive policing' to corporate sustainability management. *Journal of Cleaner Production.* 153, 673–686 https://doi.org/10.1016/j.jclepro.2016.10.175

Seele, Peter (2018a): „Let Us Not Forget: Crypto Means Secret. Cryptocurrencies as Enabler of Unethical and Illegal Business and the Question of Regulation", *Humanistic Management Journal,* 3 (1), 133–39.

Seele, Peter (2018b): *Comeback des Swiss Nummernkontos dank Bitcoin & Co.* https://insideparadeplatz.ch/2018/01/02/comeback-des-swiss-nummernkontos-dank-bitcoin-co/ (3.12.19).

Seele, Peter; Gatti, Lucia (2017): Greenwashing Revisited: In Search of a Typology and Accusation-Based Definition Incorporating Legitimacy Strategies. *Business Strategy and the Environment.* 26 (2), 239–252.

Seele, Peter; Zapf, Lucas (2017): *„Der Markt" existiert nicht – Aufklärung gegen die Marktvergötterung.* Wiesbaden: Springer Gabler.

Seele, Peter; Dierksmeier, Claus; Hofstetter, Reto and Schultz, Mario (2019): Mapping the Ethicality of Algorithmic Pricing: A Review of Dynamic and Personalized Pricing. *Journal of Business Ethics* https://doi.org/10.1007/s10551-019-04371-w

Seiwert, Martin (2015): Wenn Computer über Karrieren entscheiden. *Wirtschafts Woche,* Karriere, June 16.

Singer, Natasha (2014): Listen to Pandora, and it listens back. *New York Times*, Technophoria, Technology.
SPIEGEL (1969): Wahlwerbung: Hirnlosigkeit Heute. *DER SPIEGEL*, Deutschland, 23/69.
Stadler, Hans (2008): Lemma: Landsgemeinde. In: Stadler, Hans: *Historisches Lexikon der Schweiz, Band VII.* Basel: Schwabe.
Stalder, Felix; Mayer, Christine (2011): Der zweite Index: Suchmaschinen, Personalisierung und Überwachung. *bpb*. Abrufbar unter: http://www.bpb.de/gesellschaft/medien/politik-des-suchens/75895/der-zweite-index?p=all *(letzter Zugriff: 10.02.2016).*
Starbucks (2014): Firmengeschichte. *Firmengeschichte | Starbucks Coffee Company.pdf.* Abrufbar unter: http://www.starbucks.ch/about-us/our-heritage *(letzter Zugriff: 25.08.2015).*
Süddeutsche Zeitung (2015): Mediaten – Süddeutsche Zeitung. Abrufbar unter: http://sz-media.sueddeutsche.de/de/home/files/sz_preisliste_79.pdf *(letzter Zugriff: 10.11.2015).*
The Nielsen Company (2015): *Global Trust in Advertising.* Nielsen.
Tsukayama, Hayley (2014): How closely is Amazon's Echo listening? *The Washington Post*, The Switch, November 11.
Villiger, Simon (2014): Urnengang vom 9. Februar 2014 – Überblick der Beteiligung. Abrufbar unter: https://www.stadt-zuerich.ch/prd/de/index/statistik/publikationen-angebote/publikationen/webartikel/2014-02-13_Urnengang-vom-9-Februar-2014_Ueberblick-der-Beteiligung.html *(letzter Zugriff: 05.05.2016).*
Voytek (2012): Rides of Glory. *Uber Blog*. Abrufbar unter: https://web.archive.org/web/20140828024924/http://blog.uber.com/ridesofglory *(letzter Zugriff: 17.09.2015).*
W, Brian (2015): Sex Behind Closed Doors: Marriage, and the Invention of Privacy. http://www.annalspornographie.com/sex-behind-closed-doors-marriage-and-the-invention-of-privacy/ *(letzter Zugriff: 08.06.2015).*
Wachwitz, P. (1962): Artikel: Martin, der Heilige von Tours. In: Galling, Kurt; Campenhausen, Hans Frhr. v.; Dinkler, Erich (Hg.): *Die Religion in Geschichte und Gegenwart. Handwörterbuch für Theologie und Religionswissenschaft, Band 4.* 3. Aufl. Tübingen: J.C.B. Mohr, S. 780.
Wagenknecht, Sahra (2009): Kapitalismus heißt Krieg. *Homepage von Sahra Wagenknecht*. Abrufbar unter: http://www.sahra-wagenknecht.de/de/article/487.kapitalismus-heißt-krieg.html *(letzter Zugriff: 10.12.2015).*
Watson, Steve (2015): ‚Operation Karma Police': The British Government Spied on Everyone's Web Activity, Cell Phones. Massive GCHQ Data Bank. *Global Research*. Abrufbar unter: http://www.globalresearch.ca/operation-karma-police-the-british-government-spied-on-everyones-web-activity-massive-gchq-data-bank/5478221? *(letzter Zugriff: 13.01.2016).*

Weber, Max (1986 [1920]): Die protestantische Ethik und der Geist des Kapitalismus. In: Weber, Max: *Gesammelte Aufsätze zur Religionssoziologie*. 8. Aufl. Tübingen: J.C.B. Mohr, S. 17–206.

Wisdorff, Frank (2014): Das gefährliche Verhätscheln der Mitarbeiter. *Die Welt*. Abrufbar unter: http://www.welt.de/133487093 *(letzter Zugriff: 27.10.2015)*.

Zapf, Lucas (2014): *Die religiöse Arbeit der Marktwirtschaft: Ein religionsökonomischer Vergleich*. Baden-Baden: Nomos.

Zapf, Lucas (2018): „In Bits we trust. Glaube, Vertrauen und Währung digital", *Neue Wege*, 112 (3), 21–25.

Zapf, Lucas and Peter Seele (2015): Berechnendes Vertrauen in den blinden Glauben an den Markt. In: Baer, Josette and Wolfgang Rother (Hg.): *Vertrauen*. Basel: Schwabe, 181–99.

Ziegler, Peter-Michael (2004): Wissenschaftler stützen These von Wahlbetrug bei US-Präsidentschaftswahl. *heise online*. Abrufbar unter: http://www.heise.de/newsticker/meldung/Wissenschaftler-stuetzen-These-von-Wahlbetrug-bei-US-Praesidentschaftswahl-116587.html *(letzter Zugriff: 04.05.2016)*.

Zulauf, Daniel (2018): „Schweizer zahlen lieber mit Bargeld", Luzerner Zeitung, Wirtschaft, 31.5.

Part III

Theory of a Structural Change of the Private

4

The Private Sphere Changes: A Consequence of Digitalization

The digital age has such a far-reaching influence on the private that it requires a structural change of this private. We theorize this change in the fields of economics, politics and social affairs in the following section. The aim of the proposed theory is to provide a view of the principle of privacy that is updated to include the aspect of digitalisation.

For the creation of this theory, we summarize the previously described lifeworld symptoms of this structural change from Part III and abstract them in favor of a general theory, which we elaborate on the three levels of economy, politics and social—in which we see the greatest changes. The areas of life in which digitalisation heralded change diachronically and synchronously recur in the process, similar to a musical leitmotif, and illustrate the change in the private sphere as implied by the term 'structural change'. This can be seen, for example, in the topic of hooking up: from the analogue visit to the village festival to local speed dating and matching algorithms of international dating websites, digital methods and means of communication are changing the secret in the private sphere.

The three areas of economy, politics and social affairs provide the form in which we theoretically describe the structural change of the private sphere. Each of the three areas is divided into different crystallization points, as summarized in (Fig. 4.1).

The translation of this chapter was done with the help of artificial intelligence (machine translation by the service DeepL.com). A subsequent human revision was done primarily in terms of content.

Economic structural change of the private sector	Political and personal significance of the economy and its digitalisation (4.1.1)	Economic use of personal information (4.1.2)	The secret private as a business case: seamless products and platform capitalism (4.1.3)	The shaping of the secret private by companies (4.1.4)
Political structural change of the private sphere	Intrusion of politics into the secret private sphere of citizens (4.2.1)	Mixing politics and economics through the use of the secret private sphere (4.2.2)	Opposition to the political domination of the private spher (4.2.3)	Democratic-legislative updating of the concept of privacy, participation, transparency (4.2.4)
Social structural change of the private sphere	More exchange, less self-determination: informational heteronomy (4.3.1)	New social spaces (4.3.2)	Digital intentionality as calculating social behaviour (4.3.2.1)	"Self-policing" instead of the right to be forgotten (4.3.2.2)

Fig. 4.1 Points of crystallization of the theory of the private without secrets in economics, politics and social affairs

4.1 Economic Structural Change of the Private Sector

The economic structural change of the private, as we saw in the previous chapters, unfolds its effect on the secret private of the individual in two areas in particular:

1. Expansion of economic activities to all areas of life and thus the progressive dissolution of the boundaries between public–private–professional;
2. The growing power of intermediaries and the rise in importance of platform capitalism with access to the private sphere.

This change is based on some preconditions that are not obvious everywhere. Therefore, we first look at two systemic drivers that gave this development its shape: the social significance of the economy and the establishment of the *new economy* that accompanied digitalization.

4.1.1 Political and Personal Significance of the Economy and Its Digitalisation

One finding that runs like a leitmotif through the current presentations is the growing importance of the economy in the digital age. Once private areas, such as spending the night somewhere, celebrating, eating or looking for a partner are becoming business cases thanks to digitalisation. The primacy of the economy has taken hold of society in its full breadth and depth.

This primacy is particularly evident in the relationship between the economy and the state. It is true that political and economic functional areas are separate: on the one hand, the state monopoly on the use of force, on the other, the creation of economic resources. The hierarchy comes about through the social division of labour between these two forces: the economy produces, the state regulates what is produced and ensures an overall climate conducive to economic activity. With Habermas, we can speak of a "functional imperative of self-regulated markets" (Habermas 1998, p. 98). The economic process focuses on its participants and their decentralized decisions, the public-state organization is subordinate to this focus: for the state to function, the economy must function. Through the 'social market economy', this basic idea became a political programme, the close relationship between the modern, liberal constitutional state and the market economy is the recipe for success of modern European societies (cf. Exenberger 1997, p. 3 f.). The conviction that the market economy is the best means of preserving equality of opportunity, freedom and increasing prosperity took root. In short: the most suitable economic form for a liberal-democratic constitutional state.

Over time, politics serving the economy has become so entrenched and independent that some would like to speak of a "post-democracy" (cf. Crouch 2000). Democracy is no longer the form of political rule, and thus the people are no longer sovereign. Rather, it is an "authoritarian capitalism" as described by Peter Bloom, a "repressive logic whereby a strong capitalist sovereign is required to 'discipline' those who are economically 'irresponsible'" (Bloom 2015, p. 5). The market is stylized as a socially exaggerated model; it is idolized and thus exaggeratedly guides the fate of entire economies (cf. Seele and Zapf 2014).

This state of affairs can be traced back historically along with the expansion of economics. In the 1940s, the national economist Lionel Robbins observed that not only human activities that obviously belong to the economic sphere—such as work, trade, production—fell within the scope of economics, but also, and especially, those activities that do not belong to the economic sphere, such as leisure and hobbies. With a simple argument: all this has an influence on productivity. Not only work but also non-work determine the "material and non-material welfare" of an actor (Robbins 1945, p. 25). Robbins' successors, and especially those of the *Chicago School*, now set out to describe all aspects of life with economic theory. Gary S. Becker was probably the best known of these scholars, no area safe from his theory of rational choice. He wrote about "marriage, births, divorce, division of labor in households, prestige, and other nonmaterial behavior with the tools and framework developed for material behavior" (Becker 1991, p. IX).

Perhaps these academic considerations laid the groundwork for the expansion of the economy. Or perhaps the expansion was always inherent in the economy and merely lacked the technical means to unfold. This idea is suggested by earlier analyses of free-market economics. First of all, Karl Marx's well-known observation about capitalism as a social crystal nucleus, which his companion Friedrich Engels sums up as follows:

> By demonstrating in this way how surplus value arises and how surplus value alone can arise under the rule of the laws governing the exchange of commodities, Marx exposed the mechanism of the present capitalist mode of production and the mode of appropriation based on it, he revealed the crystal nucleus around which the whole present social order has settled (Engels 1956, p. 190).

The logic of the market economy as a "crystal core", far more than the technical structuring of exchange. Marx's broader description of the commodity character of relations (cf. Marx and Engels 1962) and the fetish character of commodities (Marx 1956) point to the early tendency of the market economy to venture far into the private sphere of its actors. As a contemporary reflection of this characteristic stands Slavoj Zizek and the concept of commodification, by means of which he describes the dissolution of boundaries in the postmodern market economy through the extension of the logic of goods to all areas of life (Zizek 2009, p. 144). In this way, Zizek refers to Guy Debord and his observation that the unconditional focus on consumption as a cultural side effect of the market economy has transformed every aspect of human life into a product. Experiences that are actually inherent to human beings (and thus pre- or uneconomic) are sold as products through advertising and mass media. Debord sees life in the market economy as a spectacle devoid of meaning, as it is only self-referential, removed from any authentic human experience (cf. Debord 1996, for a current application Heath and Potter 2005, p. 7). A striking example of *commodification* is the red-and-black Che Guevara likeness, which, long removed from the ideas of the individual behind the image, has established itself as the go to merchandise for a whole bundle of political slogans. Individual political statements become consumer goods: "Today, anyone who feels an urge to improve the world no longer pins a badge to their T-shirt, but buys a T-shirt made of organic cotton" (Uchatius 2013, p. 9).

Marx speaks of human relations, Debord of an authentic human experience, Zizek of a "fundamental structure of our society" (Zizek 2009, p. 63). All are blocked in their free development by the expansion of the economy. A return to the non-economized primordial state of experiences, relationships and societies is described as desirable. The claim to exclusivity of the economy becomes the source of the need to draw a line, to allow more than a single logic. It is, therefore, a matter of alternative logic outside the economic sphere that one can orient oneself towards (cf. Werner 2014, p. 169 f.).

For a less ideological and more pragmatic demarcation between the economy and the private sphere, it is worth taking a look at regulatory policy in this context. For example, it prohibits advertising that is explicitly directed at small children. Advertising for cigarettes or alcohol is also only permitted to a limited extent. A clear demarcation: children are protected from direct economic influence, and reference to health also prevents intoxication from being advertised too prominently (even by adults). Sensitive private areas—the offspring and health—are protected by policy from uncontrolled economic growth.

For our research question, the insight remains that the economy seems to have an inherent tendency to encroach on the shaping of life: the "economy of attention" (Franck 2003, p. 46) monopolizes its actors and displaces other topics. With the use of digital infrastructures, this monopolization of attention multiplies, as we saw from the symptoms we studied: in emailing, social networking, listening to music, streaming series, playing sports, driving cars and trains, even dating, digital information processing enables constant corporate access to the individual and his or her secrets. A comprehensive and economically driven omniscience is emerging.

The decision of a court in Queens County (USA) in September 2015 in connection with Uber exemplifies how digitalisation promotes this omniscience and the structural change of the private sphere is fuelled by the economy. The lawsuit was brought by smaller banks specialising in loans for the acquisition of taxi licences in New York. These licenses were previously auctioned off and fetched prices of up to US$700,000. Due to Uber's competition, the value of the licenses is falling (cf. Engquist 2015). Due to the lack of business, license holders also have problems servicing their loans. In the lawsuit, the plaintiffs appealed to the regulation that only licensed taxis had the right to be summoned from the roadside by customers. The Uber app would do just that, but with unlicensed drivers. The court found that an Uber ride arranged online via the app was formally unrelated to hailing a taxi from the curb. In addition, the court recognized the plaintiffs' intention to use their lawsuit to compensate for a dwindling customer base. The court disagreed, clarifying that "It is not the court's function to adjust the competing political and economic interests disturbed by the introduction of Uber type apps" (Weiss 2015, p. 6). The court, one might interpret this comment, sees Uber as part of an evolutionary market process. In the absence of legislation to the contrary, the court does not interfere with it. The autonomy of the economy—even if it is perceived as unjust—remains intact so that digital innovations can unfold their economic potential undisturbed.

4.1.2 Economic Use of Personal Information

At the heart of digital entrepreneurial activities is the processing of personal information. Uber, to stay with the example, collects location, usage, personal and technical information thanks to numerous sensors on mobile phones—the app cannot even be opened without access to these. And it does so in advance, during the ride and after the ride, and even when the app is not even open. Always tied to the collection is the mantra: we need this data "[To] provide or improve our services" (Singer and Isaac 2015). The concept, in Uber, as in other similar companies, is based on the large-scale, automated and networked collection and use of personal data. On the one hand, and ostensibly, as a service to users—organizing a cheap taxi service or operating a homepage that serves as a social network. On the other hand, the information serves as a source with which to create a map of social interaction. The insights can be lucratively resold, separated from the context in which the data originated (cf. Andrejevic 2016, p. 32).

This double-entry bookkeeping of information processing is characteristic of the offerings of the *New Economy*. The digital infrastructure is available to the user free of charge or very cheaply, in return for which data is obtained during use, processed further and advertising profiles created. This enables access to "the real-time flow of our everyday lives with the aim of directly influencing and changing our behaviour and turning it into a business" (Zuboff 2016).[1] Products that exactly match the respective profile are selected

---

[1] For a quarter of a century, Zuboff has been concerned with the question of how various forms of surveillance affect the lives of the monitored. She became known with her book *In the Age of the Smart Machine* (1984), in which, at the beginning of the digitization of the workplace, she pointed out the accompanying possibilities for control and the risks, describing it as a new form of panopticon.

and communicated in channels that are characterized by a particular proximity to the user (cf. Hill 2015a, b): the uncomfortable shoes complained about in the email to one's partner are acknowledged within the mail application with an ad for comfortable sandals. This works. A less subtle example: the anti-virus manufacturer AVG sells the browsing history of its customers to advertising partners. In exchange, the user gets a free anti-virus program (Softpedia.com 2015). For the user, the currency in the New Economy is no longer money, but personal information: age, gender, place of residence, socio-economic status, current interests. A barter transaction, information for the product. The collection and use of information are made transparent in long terms and conditions. In most cases, however, the user clicks the "Read and accept" button at record speed in order to finally access his e-mails or start the virus scan.[2]

The business model outlined requires users to be open with their information. This is systematically promoted by the providers. For example, Facebook sees it as its mission to get people (meaning Facebook users) used to revealing their entire personality online as *an* 'authentic identity'. All facets of the personal should be accessible without hesitation on Facebook. According to Facebook board member Sandberg, this is not far off:

> Expressing our authentic identity will become even more pervasive in the coming year. Profiles will no longer be outlines, but detailed self-portraits of who we really are, including the books we read, the music we listen to, the distances we run, the places we travel, the causes we support, the videos of cats we laugh at, our likes and our links. And, yes, this shift to authenticity will take getting used to and will elicit cries about lost privacy. But people will increasingly recognise the benefits of such expression (Sandberg 2011).

The company's vision of 'authenticity' fits into the business model outlined above because 'authentic identity' is a necessary virtue of customers at both levels of the information business—the social network operation and the advertising business. An anonymous alias has no friends and cannot be influenced by advertising.

---

[2] The question arises as to the effectiveness of this customer information. The prevailing logic becomes even clearer in the now mandatory notices in Europe that websites set cookies. The user is informed, but can only press "OK". A refusal or "opt-out" is not made possible. From a privacy perspective, a completely different approach would be desirable, namely an "opt-in" for all personality-relevant online structures: at the beginning of use, explicit consent would have to be given regarding the type and scope of data use. The fact that the service cannot be used under certain circumstances if consent is not given remains unaffected. However, such an opt-in would be helpful in terms of the comprehensive perception of data use.

If we summarize the two preconditions of the economic structural transformation of the private sphere, the overall result is a compelling economy that is highly present in the reality of life and whose business model is based on aggregating, processing and selling personal information. An economy that, supported by technological innovations and their social anchoring, makes large-scale use of the private and secret private of its actors. This dissolves the boundary between the private and the economic—in the language of the providers: better and better offers are being created for more and more areas of human life (cf. Schulz 2015). In this economic expansion and use of information, the economic structural transformation of the private takes place.

4.1.3 The Secret Private as a Business Case: Seamless Products and Platform Capitalism

The fact that the economy is not only interested in the private, but especially in the secret private, becomes apparent by observing the symptom of advertising and recommendation and its three-steps towards digitalization: the classic newspaper advertisement as a communicative one-way street—use of the secret private is excluded. Then direct marketing based on the link with the personal environment and thus the first steps in the direction of the secret private until then—the last stage of expansion—the individualised personal address of the customer based on the current interests becomes possible on the basis of the digitally collected information. This is exemplified by the case of the woman who received advertising recommendations for baby products based on her shopping behavior before she even knew she was pregnant.

The secret-private information becomes a business case. It is promising for companies to personally address individual customers and their needs by analyzing individual data sets. The individual is at the center of economic attention as one employee of the music streamer Pandora puts it: "The advantage of using our in-house data is that we have it down to the individual level, to the specific person who is using Pandora," (Singer 2014, p. 2) *Mining* is deepening: companies are no longer just collecting metadata, such as location and movement information, credit card data and analyzing it anonymously, but identifying individual customers from this information: cultural background, religion, political views. Companies get close to the needs of individual customers with this knowledge: a real-time market analysis with an individual customer approach.

For successful individualization, the user experience must be organized *seamlessly* and without friction. It must not be interrupted, for example, by changing the end device, leaving the home or the website, switching from email to video chat. The product should accompany the customer smoothly. Accordingly, products are needed that are physically close to the customer. In the trouser pocket or on the arm, *wearables make* this goal possible: bio-measurements, video and sound recordings and thus a comprehensive informational infrastructure directly on the customer. The various offers on the devices are linked to form *all-in-one solutions*: chat, social network, email, search engine, tablet, PC, phone, watch, all in one. For perfected examples of such a frictionless product range, just look at offerings from Apple or Google. Seamless integration allows the products to fit into the lives of their users. Communication, monitoring of vital functions, work, leisure, mobility, there is an offer for every situation, including feedback on product use to the provider.

The interaction, which is deeply integrated into everyday life and is taken for granted, guides the customer's perception and subsequently his purchasing decisions. For example, through search suggestions, 'you might like this', health and nutrition tips, media tips, travel tips, job offers. The incentive for companies is to capture the individual customer via their digital profile so authentically that the virtual image can be used to communicate seamlessly with the customer's real-life perceptions and decisions. In the future, even more extensively than today. For example, with glasses that measure the wearer's field of vision and display information according to the measurement—e.g. similar products when walking past a shop window or the link to buy a piece of music when the radio is playing. The technology for this already exists. It comes from military research and was designed for situation assessment in the field: to provide the soldier with information that makes sense in order to make decisions in confusing situations. As it turns out, this also works in the battle for the customer's favor:

> [The] device might be used to tag consumer items or even specialty shops that catch your fancy as you walk down a city street. Just a quick glance at a dress in a window, for instance, might elicit a neural firing pattern sufficient to register with the system. A program could then offer up nearby stores selling similar items or shops you might want to investigate. [...] There's so much information to explore and digest, how do you make it useful to a person at a given time? We can make it unobtrusive and tag things as you move through your environment (Daley et al. 2011, p. 38 f.).

Such a device blurs personal and economic realms by directly influencing the perception of its wearer. Thus, even the furtive glance is no longer withdrawn from economic use: *according to your field of vision, you are particularly interested in the opposite sex right now*. A discreet escort service in the vicinity appears. The interaction between companies and customers creates its reality, based on secret and private information and shaped by economic interests.

Digitalization makes it possible to support business concepts broadly, decentralized and accessible to everyone. In this strategic expansion, the resources of the participants are central, while the task of the entrepreneur is to provide a digital infrastructure and information collection. This is referred to as platform capitalism (cf. Lobo 2014). A platform functions outside the fixed definitions of providers and demanders: providers can be demanders at the same time, and vice versa. The offers on platforms are flexible, the barriers to entry conceivably low. They can be registered quickly, are available at any time and from anywhere, and are usually free of charge. Even if the sharing offers to be low-threshold: behind them are now huge corporations, billion-dollar companies that use private resources to make money (cf. NDR 2015). Real power relations are most likely to come to light when conflicts arise. When taxi drivers in Milan and Paris threaten Uber drivers when Airbnb is to be banned in Barcelona. Then, the companies speak out, with a concentrated legal armada from overseas, and make it clear that first the market will be rolled up and then the remaining issues will be clarified. Aggressive strategy in the look of private sharing, with far-reaching consequences:

> Quite soon, within the next two decades or so, less than 50 percent of people will have jobs for which they have been trained (i.e. agriculture, Industry or services). Even highly skilled jobs will be at risk (Helbing 2015, p. 16).

The economic structural change of the private sector changes not only the sector itself but also the entire labour market through the outsourcing of services to private actors.

The economic use of the private sphere is not without influence. Wearables, seamless design, platform offers are getting so close to the individual that they are beginning to have an influence on them. As the example with the data glasses shows: now, it is no longer about knowing what the customer is currently perceiving, but influencing his perception. An intensification of the interaction between customers and companies described above.

4.1.4 The Shaping of the Secret Private by Companies

How the economic *use of* private information results in the *shaping of* the private can be seen in the symptom of dating examined here: from the village disco to speed dating to the algorithm that selects a suitable partner, more private information is passed on to third parties at every step. If it is only the visible person when visiting a disco, speed dating already requires membership and an associated profile. In online dating, pages and pages of profile information are filled in, pictures are selected and the partner search algorithm is set in motion based on this. Based on the data provided, the dating platform makes connections that the partner-seeker would not be able to come up with themselves (cf. Bridle 2014). The volunteered data and the digital additions to this data provide the user with clues for standardised, comparability-producing measurement of his or her self. In terms of dating, this 'surveying'—which harmlessly emerges as a *matching score*—is linked to a specific view of the potential partner. It becomes relative: 8/10 requirements achieved, 20% more education, but deductions in appearance. As a further consequence, the found counterpart is a position in the measurement grid—there are worse matches, but perhaps also a few better ones. Logical consequence: the profile remains online, even if one has got involved with a match. A better match might come along.[3] This is where the transition from use to shaping arises: the digital offer influences the perception of one's relationship. The corporate shaping of the secret private shows up on two levels. Firstly, in the systematisation and rationalisation of the self, in that profile information is collected (Who am I?), expectations are explicitly formulated (Who do I want?) and thus measurability is assigned to one's status and that of the partner. As a result, a virtual market value of the persons involved is created.

There are a number of consequences for the secret private. First, the boundary between the economy and the private sphere continues to blur. With the use of digital infrastructures, the actor is at no point 'left alone' by economic considerations—one of the core requirements of privacy. Through the potentially constant use and analysis of the digital self, the actor is deprived of the decision as to which part of their utterances and actions should be understood in relation to the economy, by whom the information is used and what is done with it. A sample of how this abstract change manifests? The topic of nutrition. An internet search for the benefits of a vegetarian diet and a few

---

[3] This is not to say that this behavior is not observed in offline dating. However, the digital infrastructure using the secret private supports and promotes this behavior by processing the necessary information (and the view of the possibility: x-million singles!) and finding 'the perfect match' with the promise of a scientific procedure.

posts on relevant forums can result in samples of tasty meat snacks landing in the mail of a renegade carnivore a few days later. The sender is a large meat manufacturer who received the address from the search engine operator. The meat tastes quite good after all, so vegetarianism is reconsidered (cf. Morozov 2014, p. 400). The example shows the problem of the economic shaping of the secret private: content is individualized by analyzing user behavior (internet search, forum entry), comparing it with similar behavior patterns (ending meat consumption) and thereupon steering it according to a certain interest (sales of the meat producer). An external player guided by his interests constructs the consumer's perception in this way. In analogy to 'predictive policing'—the consequence of Big Data analysis at the state level—one could speak of 'predictive marketing'. This practice seems problematic because it limits choice. Targeting is done on the basis of prior preferences, and the countervailing information is hidden. This plays against individual reflection while the consumer thinks they are making an independent decision. The private is normalized under the guiding principles of the economy, an expression of the ongoing marketing and subcontracting of the individual via the digital (cf. Grassegger 2014a, b; Sennett 2010).

Behaviour that seeks to evade these processes is "perceived, disapproved of and sanctioned in the digital public sphere as an element of social disorder" (Crueger 2013, p. 23). The constant awareness of being subjected to economic interest formation creates its binding force, a new type of bourgeois virtue (cf. McCloskey 2006, p. 507). The individual internalization of free-market logic updates itself around the latest methods of marketing: accept that the economy knows everything about you, uses you and shapes you. In the economic shaping of the secret private, we see its abolition.

4.2 Political Structural Change of the Private Sector

In Habermas, we could read: "The public sphere is limited to public violence" (Habermas 1990, p. 90). Politics is part of this public violence—and thus, according to Habermas, not part of the private. Following our accounts of digitalisation and its impact on the secret private, it is clear that the strict demarcation between public policy and the private life of the citizen no longer exists. In the digital age, the state's access to the citizen becomes more immediate, extensive and inescapable. The political is no longer the main function of the public, but directly accesses the realm of the private, which changes the individual's position on politics. These new relations of the secret private and the connections to political power are explored in the following section on the political structural transformation of the private.

The central theme in the reorganization of these power relations is the technical possibilities and digital infrastructures with which state actors fulfill their political mandate: fighting terror thanks to email control, lowering healthcare costs thanks to access to the jogging app, ending money laundering thanks to the switch to electronic payment. The oversized nose lends itself as a leitmotif of the political structural transformation of the private sphere: the 'snooping state' invading areas that had been denied to it in the analogue republic. Accordingly, we begin by describing the new possibilities of surveillance as the precondition and strongest feature of this structural change. In particular, we describe the entanglement between economy and politics in this context as a structuring characteristic of this domination. As an antithesis to this development, we describe the opposition to this political domination of the secret private sphere.

4.2.1 Intrusion of Politics into the Secret Private Sphere of Citizens

In 2013, a joke made the rounds on the internet that went something like this: *I met Barack Obama and told him, 'My dad thinks you're spying on all of us.' The president replied, 'He's not your dad.'* In a concentrated form, this joke reflects the actors, perceptions, and potential consequences of current government surveillance activities:

- *Actors:* Since Edward Snowden's revelations, it has been clear that the US administration has the world's largest infrastructure for spying on the secret private (cf. Bamford 2012). Therefore, in the joke, Obama provides the answer.
- *Perception:* There is a critical awareness among the population that information on the secret private is collected on a large scale (cf. Levison 2014)—hence the child's demand.
- *Possible consequences:* And the data, even when promised otherwise, will of course be used by various agencies to spy on their citizens (cf. Savage 2016). So Obama knows who the real father of the child is. So it shouldn't come as a surprise when an American genealogy site is forced by the government to hand over DNA files that were voluntarily submitted by a customer for the purpose of discovering his genealogy, in order to facilitate a criminal

prosecution (cf. Hill 2015a, b). The mobile phone surveillance established in counter-terrorism is also to be used for other offences, such as tax evasion. Because otherwise, every citizen would have a Swiss account in his pocket, as the American president claimed (Reuters 2016; Savage 2016).

Whereas in the past the Stasi was the epitome of surveillance and spying, and the GDR the perfidious market leader of innovative shadowing methods and disregard for the secret private sphere of its citizens, in the digital age this honour is apparently being bestowed on the USA.

The change in the secret private sphere through digitalisation is particularly evident in the symptom area of state surveillance. It all began with the village policeman, who knows 'his friends' and performs a social function in the clearly defined area of his village. Supported by the village eye, the watchful look out of the window. But the front door marks the end and the neighbour is not always there to look. With the widespread establishment of video surveillance, state surveillance and its influence on the secret private sphere changed on two levels: the cameras are always on, filming regardless of the people recorded or whether there is problematic behavior or not. And the recordings are stored. Digital memory leads to the reproducibility of 'who', 'where' and 'when' independent of suspicion, only due to one's presence at a location that is video-surveilled. Digitisation consolidates and extends this paradigm shift that began with camera surveillance. The recording of digital communications covers the secret private more comprehensively. The video was only able to capture the mere presence in public space. Now, surveillance penetrates directly into the person, becoming latent and richer. In addition to 'who', 'where' and 'when', there is now the 'what', perhaps even the 'why' of individual actions.

The digital simplifies surveillance: collected information is easy to send, copy and search. And the vast majority of the population, even if they use electronic media and communication methods with restraint, independently contribute to generating masses of data sets. Via mobile telephony, the Internet, search engines and apps, and not least in dealings with the state itself—electronic tax returns, biometric passports, the virtual town hall—personal information is being funneled to the state through cyberspace. Digital infrastructure overcomes the boundaries between the state, citizens and individual companies (cf. Schneier 2015). In the process, the secret private dissolved by the state creates information asymmetry. Surveillance is only effective if the information collected is kept secret from its originators—the individuals under surveillance. Accordingly, the citizen is not informed about

4 The Private Sphere Changes: A Consequence of Digitalization

what secrets are stored about him by secret services. A new kind of secret private—without the right of access of the originator of the secret.

The digitalised state apparatus generates a new form of transparency through this information asymmetry (cf. Hansen and Flyverbom 2014): instead of a transparent state, a transparent citizen emerges who, because his or her self is being monitored, lives in constant uncertainty about whether his or her actions will be monitored or the information collected will be used against him or her at some point. Once the secret private is abolished, it is only a matter of policy to use the knowledge to prevent serious cuts to freedom and civil rights. The suspicionless collection of data and its complex scope gives the impression that the state has access to a compelling picture of those under surveillance: the collection of personal contacts, credit card data, movement profiles and mobile phone data gives rise to a detailed picture of individual life realities.

However, the combination of communication and metadata does not necessarily paint a realistic picture (cf. Gonzalez 2012). Technical surveillance and its legal exploitation are thus tantamount to prejudgement or, to put it more theoretically: through the selection of methods and capabilities, the state forms for itself an image of the secret private sphere of its citizens that is not necessarily congruent with the reality of life.

A particular feature of the political structural change of the private sphere is that it takes place with little resistance. The citizens under blanket surveillance seem to have resigned themselves to it, and this situation is met with a mixture of conviction that it is already the right thing to do and a lack of interest: "The technological advancements crept up without much reflection, and the public was simply not engaged nor concerned about lofty privacy matters" (Heumann et al. 2016), it is said in relation to the omnipresence of video surveillance. It is true that there are groups that draw attention to the dangers, data protection officers, social scientists, artists, who repeatedly raise the issue. However, these activities remain in a minority position in the background for at least two reasons:

- First, the creation of public security is the concern of a majority of the voting population: the deal of 'security at the expense of privacy' is supported by the majority. As read in Thomas Hobbes *Leviathan* (1651), fear of death and convenience are the starting points of the formation of a polity. Politics is attempting to address both of these points with its digitized intrusion into citizens' privacy. It is successful in doing so because citizens naturally seek interaction with the technologies underlying surveillance. This may well be seen as a modern form of convenience. An inherent desire for

security in the polity seems to encourage the abolition of the secret private sphere.
- Second, and probably more serious than the political will of citizens to be monitored, is the painlessness of the invasion of privacy. Governmental incursions into the private sphere are invisible and unnoticeable. You do not know if, when, what. As long as the police do not kick down doors, take away, imprison, accuse, the abolition of the secret private is an abstract problem (cf. Schneider 2015, p. 4).

Increasing interaction with technical devices and access possibilities to connected sensors mean that state surveillance activities can comprehensively take place close to people. Under these conditions, surveillance takes place across the board, in a general manner rather than for a specific reason. Citizens grant the constitutional state these measures and thus access to their secret private sphere.

4.2.2 Mixing Politics and Economics Through the Use of the Secret Private Sphere

Politics and business go hand in hand in the abolition of the secret private sphere. On the one hand, politics acts as a beneficiary, on the other as a customer. For business, this cooperation is not a social end in itself. Rather, the Hobbesian formula of convenience and security and the digital revolution as its solution promise tangible profits, the secret private of citizens as a business model. The mutual use of the emerging pool of knowledge seems expedient for business as well as for politics, each with different interests (cf. Schneider 2015, p. 5).

In particular, the state is taking advantage of the innovations in the area of *advertising and recommendation:* combining personal data and user behavior to create profiles. With this infrastructure, the state only has to get to the sources, for which it legally obliges telecommunications companies to cooperate. Take Bude in Cornwall, for example: here GCHQ, the Britische Geheimdienst (British Secret Intelligence Service), literally sits on the overseas cables coming out of the sea, connecting Europe's and America's internet communications, and cuts in cf. Hopkins and Borger 2013).

Things are somewhat more subtle in the other direction of the economic-political mixed use of the secret private sphere when the state is supported in its core tasks by corresponding companies. A current Google project is converting old public telephones in New York into high-speed hot spots (cf.

Lapowski 2015). Within the framework of such public-private partnerships, market-leading companies act as infrastructure partners for cities and municipalities that cannot afford the expensive expansion of digital networks. Cost-neutral for the taxpayer, the companies step in and simultaneously sit at the source of the resulting data. Companies are also taking over government tasks in less visible areas. Another example from the same provider: Google wants to display customized results when certain search terms are entered (e.g. the search for participation opportunities with terrorist organizations), spread disinformation, and display free offers according to exit organizations (cf. Barrett 2016). Other companies are also proactively participating in the fight against terror, for example, by identifying terrorists on social networks. Personal data, which is collected anyway to optimize the offer, can be used to combat deviant behavior without much effort. If necessary, companies adapt the corresponding algorithms, one is interested in a good relationship with the state. Corresponding information is passed on to authorities voluntarily, i.e. initially without a legal basis. Instead of an obligation to hand over information against a court order, a debt to pay arises that is worked out by the company on the basis of independent criteria (cf. Yadron and Wong 2016).

Overall, the state can outsource expensive infrastructures, but still, use them for surveillance. *Win-win is what* you call such a constellation. Clearer *lots,* however, for informational self-determination and an infrastructure that is neutral towards user data. The state is now sharing power with certain companies (cf. Helbing and Pournaras 2015).

In this structure, the state is not only a beneficiary but also a customer. The market for surveillance is huge and states depend on the know-how of companies. In China, Cisco and other Western companies are helping the government to develop a nationwide infrastructure for the digital analysis of camera surveillance (cf. Langfitt 2013; Chao 2011). Not only on the executive level but also on the constitutive level, corresponding cooperations can be observed, as evidenced be the symptom area 'election and political advertising'. Politicians use Big Data-enabled systems to fine-tune their campaigns in a targeted, effective and efficient way. The election poster appears as an analogue relic, while the actual political persuasion takes place in a personalised way on the voter's individual screen.

In the USA, the corresponding movements between business and government, which usually take place behind closed doors, can be measured against a publicly accessible source: the exchange of personnel. There is a remarkable fluctuation between high-ranking technicians from Californian Internet companies and equally high-ranking representatives of the administration from Washington. In this context, the media already speak of a revolving door between the White House and Silicon Valley that rarely stands still (cf. Vincent 2016).

The term *post-democracy* was coined by the British political scientist Colin Crouch to describe the entanglement between politics and the economy, between benefiting and consuming (cf. Crouch 2000). This describes politics that is permeated by corporate-market interests. The structural change of the private sphere and this post-democracy are linked by the central role of domination over information and communication channels. Crouch describes the beginning of large-scale communication control as a means of (democratic) politics along with the progress of the US advertising industry in the 1950s. Strategies used for consumer products also worked for political content. The methods for both areas were the same,

> [...] extrapolating from the innovations of the advertising industry and making themselves as analogous as possible to the business of selling products so that they could reap maximum advantage from the new techniques (Crouch 2000, p. 10).

Politicians use elaborate marketing methods to persuade people, now firmly established in political discourse: party programmes are understood as a *product*; they are used to sell a *message*. Politics chose the short, catchy form of the advertising message as its communicative template. With consequences for the content: no elaborate, evidence-based discourses, but clear, catchy, simple phrases. Accordingly, access to knowledge about what information is currently in demand and relevant among the population is crucial for shaping policy.

We are experiencing the political structural change of the private sphere as an expansion of political claims to use the secret private information of citizens: the state uses the collected information to realize political goals, for surveillance and communication optimization. The secret private sphere is made game for day-to-day political business, such as election campaigns. The infrastructures necessary for the collection are entrepreneurial in nature—the interconnections between politics and economics are profound.

4.2.3 Opposition to the Political Domination of the Private Sphere

The structural informational asymmetry created by the political structural change of privacy provokes new areas of opposition. As an object of civic engagement, *privacy* has been known to a broad public since the 1980s. At

4 The Private Sphere Changes: A Consequence of Digitalization

that time, the computer-assisted analysis of citizens' data in the context of the census was perceived critically, and interest groups formed to organize resistance to the census. The concept of informational self-determination dates back to this time. Today's activities on the topic have expanded and are not only aimed at limiting government access to citizens' information. The scene is concerned with encryption technologies, with new types of ownership and *public domain*, with innovative forms of cooperation and community. Old categories of analogue life are being reworked. The Internet is a free space whose anarchic flair overcomes the limitations of the analog world, without spatial power structures and entrenched social relations. The preamble of the Chaos Computer Club, already versed in the subject since the early 1980s, reflects this utopian concern:

> The development towards an information society requires a new human right to worldwide, unhindered communication. The Chaos Computer Club is a galactic community of living beings, independent of age, gender and descent as well as social position, which campaigns for freedom of information across borders and deals with the effects of technologies on society as well as on individual living beings and promotes knowledge about this development (CCC 2016).

The CCC considers the potential to be so great that it does not even want to limit itself to Earth as a sphere of action, but sees itself as a "galactic community". The political nature of the club is clearly stated with the demand for a new human right.

However, the plan is easier formulated than implemented. The existing, highly complex systems are difficult to control, even under the rule of law. For the most part, citizens have no choice but to trust the state not to use the collected information as it pleases, with no oversight. In addition, the systems are prone to error. The mere fact that Edward Snowden was able to steal sensitive information on such a scale shows that the data cannot be stored completely securely. Further revelations—also driven by motives other than those of Snowden, such as revenge or frustration—could make data public that would make the lives of the monitored very unpleasant (cf. Levison 2014). Even advocates of a laissez-faire policy on data protection, who like to claim that they have nothing to hide, have to face this danger: it is not about hiding minor and major scams from a trusted state. It is about potential access to one's entire digital life by uncontrollable third parties.

Perhaps nourished by this sword of Damocles of the data leak, self-censorship is particularly evident in political surveillance. Those who already make themselves suspicious because they deviate from the mainstream keep their trenchant opinions to themselves, sheering back into the fold. The mere knowledge of the possibility of surveillance leads to the panoptic effect and normalizes discourse (cf. Garland 2001). This normalization becomes virulent because digital surveillance is unbounded. Before the digital age, state encroachment on privacy was a selective measure to punish deviant behavior. Today, without restraint, every citizen is constantly monitored—being 'left alone' by the state has become the exception. As the example of data retention shows, this is a daily reality. Corresponding laws stipulate that providers must record the metadata of *all* telecommunications and store it for months. In Switzerland, for example, all mobile phone calls have been recorded for 6 months since 2004.

- The start and finish number
- SIM, IMSI and IMEI (card, subscriber and device number)
- Location during connection
- Date, time, duration of the connection

stored (cf. Kire 2015). In other words, communication unobserved by the state is no longer possible. Without this being met with the widespread objection: "What is the harm done when privacy is violated?" (Jarvis 2011, p. 20).

However, awareness of the problem seems to be so low that politicians see a need for action (admittedly in areas where they do not benefit from it): the permissive sharing of photos of one's children on social networks recently led to a corresponding warning by the police in Germany. Careless parents were made aware that naked photos of their offspring, provided with location and time information, could possibly have a detrimental effect on development and safety in the global and ever lasting digital space (cf. Dahlmann 2015).

Non-governmental organisations are attempting to raise awareness of the problem with sometimes aggressive countermeasures. This is the case with the following sticker, which can be found throughout Switzerland (Fig. 4.2).

Against a black background, one sees an index finger vigorously poking into the lens of a camera. Underneath the text: "Lo stato ci osserva. Caviamogli gli occhi!". In English, "The state is watching us, let's poke its eyes out!" Similar statements can be found in relation to the corporate use of the secret private and its lack of state regulation: "Google is Watching you!" is written on graffiti, and the image of the 'Big Brother' of 1984 and its dystopian conception of society emerges. Civil disobedience is used to generate attention for surveillance mechanisms (cf. *World Under Watch* 2012).

4 The Private Sphere Changes: A Consequence of Digitalization

Fig. 4.2 Declaration of war on the surveillance state. Own photograph clz, 27.04.2016, Lugano/TI

4.2.4 Democratic-Legislative Updating of the Concept of Privacy

In addition to civic activity on the subject, democratic institutions themselves are also dealing with privacy. Legislators are grappling with the need to update the understanding of privacy: the old world of law collides with the latest developments.

Liberal democratic states under the rule of law grant their citizens the right to a secret private sphere with regard to communication channels (e.g. secrecy of correspondence and telephone) and their physical private sphere (e.g. inviolability one's own home). However, the concrete manifestations of this privacy seem to have been altered by digitalisation; accordingly, we observe a certain legal uncertainty with regard to the concept. Given the profound structural change, this may not be surprising. A striking expression of this uncertainty is the resistance of individual institutions of the European Union against American companies, such as Facebook and Google. Under the terms "right to digital oblivion" and the concern to enforce European data protection principles, there is a fight against the normative power of the factual, with which the companies curtail the informational self-determination of their customers (cf. Brössler 2015). According to the motto: "We can just stand back and watch; 'the Internet' will take care of itself—and us. If your privacy disappears in the process, this is simply what the Internet gods wanted all along" (Morozov 2014, p. 43). Globalized popular culture and its use of the digital collide with old-world notions of privacy. The paradox of advocating restrictive use of personal data in the citizen-corporate relationship, while at the same time pursuing the complete erosion of the secret private in the citizen-government relationship, seems inherent to politics on this issue. In any case, politics must reassess the legal valuation of that space to whose integrity the citizen has a right. The concept of privacy must be supplemented by digital reality.

The political structural change of the private sphere leads to a reassessment of political participation in addition to the confrontation with entrepreneurial interests and a new concept of privacy. As was made clear in the symptom area on networking, digital communication structures lead to an expansion of the audience, new possibilities for cooperation emerge. With concepts, such as *liquid democracy*, mixed forms of participation between representative and direct democracy can be realized by digitally supporting decision-making. Everyone can contribute to the topics that interest them. The rest is done by elected representatives. Political participation is no longer limited to the act of voting but becomes part of private action in which participation is realized with the help of digital infrastructures. This reinterprets digital infrastructures. Their current economic-political-monitoring function becomes a participatory, emancipating one.

This change means freeing oneself from the confining but self-inflicted domination of the secret private sphere and, according to the sociologist Helbing, therefore, has a character that is as enlightening as it is imperative for a free society:

> It is therefore time for an Enlightenment 2.0, leading to a Democracy 2.0, based on digital self-determination. This requires democratic technologies: information systems that are compatible with democratic principles—otherwise they will destroy our society (Helbing et al. 2016, p. 52).

The political and economic use of digitalization is thus transformed from an unpleasant transitional phenomenon into a socially beneficial novelty.

Another new area concerns transparency, a central concept for the political use of digital infrastructures. First of all, there is a negatively perceived form of this transparency: the transparent citizen. The digitally supported, automated and remotely controlled intrusion into the secret private creates asymmetries between citizen and state. It creates an imbalance of information and resources to use that information. Opposition to the blanket, systematic analysis of the secret private, the infrastructure of which is either state-owned or corporate, also means opposition to this form of transparency. In addition to the disclosure and critique of those purposes underlying surveillance, the reduction of these asymmetries is a demand of the opposition.

At the same time, innovative forms of transparency are the achievement and driving force of digitalization. The success of Web 2.0 is based on sharing, on publishing individual realities of life, and precisely not on being segregated, private. On the contrary, with this transparency, privacy should exist at most as indifference: "When people don't care enough to look, then privacy will be restored. This is a common hope in the 'transparency' movement" (Lanier 2014, p. 331). This structural transparency and publicness are seen as a successful pattern for reorienting politics (cf. Jarvis 2011). With, it is argued, many positive consequences: information about political action, failure or success spreads rapidly and is difficult to control. This also applies to the information on the integrity of public figures and their environment. A transparent policy, i.e. one whose means and ends are comprehensible, is getting a little closer.

With the awareness of transparency on both sides, on the part of the citizen and the state, a tension arises between the private claim to intransparency and public-political transparency that is perceived as desirable. The need to distinguish between these two forms of transparency in discourse and political work seems vital for the persistence of a secret private and a democracy-promoting use of the digital.

Overall, the political process lags behind economic and social developments in updating notions of privacy, participation and transparency. When Google is fought, electronic voting is prevented, or transparency is understood mainly in relation to the actions of the citizen, this reveals a tension between an outdated understanding of privacy and its legal legitimacy and the reality of life. The secrecy of correspondence and the inviolability of the home are notions of analog self-knowledge. They collide with the ways and means of modern forms of life, of e-mail and smart TV.

4.3 Social Structural Change of the Private Sphere

The social structural change of the private changes interpersonal exchanges. We capture this change on the individual level and observe in the following theory section how the abolition of the secret private here shapes the social structural change of the private.

While the collective use of digital infrastructures is often associated with post-social, post-humanist or even autistic behaviour (cf. e.g. Botz-Bornstein 2015, p. 47), we argue in a different direction and initially view digital social behaviour as a way of expanding social exchange in various directions (Sect. 4.3.1). First quantitatively, as the number of participants in digital exchanges increases rapidly as transaction costs fall. Then, qualitatively, as the exchange expands and personal roles become more complicated, even leading to a loss of informational self-determination.

In this changed structure, new social spaces are emerging in which the secret private sphere is directly threatened. Social behaviour is influenced in the direction of digital intentionality and latent self-monitoring (Sect. 4.3.2).

4.3.1 More Exchange, Less Self-Determination: Informational Heteronomy

The expansion of social exchange is related to the mass diffusion of the underlying technologies. Participating in digital social exchange becomes must, reinforced by the positive perception of these infrastructures: the Internet as a haven of progress, better distribution and communication. And all this at virtually zero cost. Mass use is easy when the transaction costs of social exchange tend towards zero. In 1999, around 10 million chargeable text messages were sent per day in Germany. In 2015, it is still around 40 million. But in addition, 670 million free WhatsApp messages (cf. statista 2016). Almost 70 times more low-threshold free exchanges within 16 years.

The circle of participants in digital social interaction can be expanded at will without transaction costs. Information can be published quickly and accessed from anywhere. If organising a parents' evening, for example, was previously a arduous affair—written invitation, changes of date via telephone chain: call to the alphabetically next person on the class list—the smartphone infrastructure has greatly simplified matters. Every elementary school class now has not one, but several chat groups: teachers, parents, teachers and parents, students with each other. No information is lost. Problems are discussed and dealt with in real time (cf. Wollenhöfer 2014). Whoever, in school, but also everywhere else would like to enter into an exchange, will easily address a variety of interested parties. Due to the enlarged audience, the platforms of social exchange are differentiating themselves, a variety of new forms of communication are emerging, with which exchange is possible at any moment from anywhere: writing, telephoning or seeing each other via a video chat.

The circle of participants is growing, as are the possibilities for communication. The roles within these social contexts are thus becoming less defined. In psychology, a 'social role' describes the adaptation of individual behaviour according to social position. This results in different roles depending on the environment: among friends, one is a friend, among parents a child, among business partners a customer. Different contexts demand different reactions. The digital age leads to a multiplication of these roles. The digital self can be reinvented without much effort and can act differently depending on the digital context. A clear expression of these multiple digital roles, which are independent of analogue roles, is the crude way of dealing in forums or reader comments. The tone here is at best unpleasantly opinionated, at worst personally insulting. Cases are regularly reported from digital social networks in which users are subjected to sexist, racist or murderous tirades from indiscriminate rabble-rousers—in response to mostly harmless comments (cf. e.g. Stokowski 2016). When acting digitally publicly from the private sphere, the harsh or brutalized tone is common. The digital self finds itself in a role that it has difficulty acting out in the analog social context: unfiltered, direct, personal, and at the same time distant from the unknown addressee.

The multiplied roles of digital social behavior cannot always be separated. In certain contexts, they overlap, even with analogue roles. This is illustrated by the example of working. Here, the potential for mixing and consequently role conflicts are already evident in the analogue. The non-working private person is constitutively part of the working person: resting from work, whether after 8 h or 6 days of labor, is inevitable for the successful resumption of work. The working person, thus, already combines the two social roles of *work* and *private life* without a digital infrastructure. The novelty of digitalization is the dissociation of these roles. In the pre-digital[4] work environment, a clear change of roles is made when leaving the office, with the saturated punch of the time clock. This clarity disappears with the use of digital communication options. The crystallisation point of this development is the smartphone tracking of employees by their boss after hours or the work email account on the private computer. The WhatsApp parents' group is also particularly happy to be active after 7 p.m., which is unlikely to please the exhausted teacher. Similarly, the role overlap works in the other direction: a circular email from the office is used to plan the football evening. Through digitalisation, the professional and private person expand into the respective other areas of life—simultaneously as a source of stress and as a positive compensatory effect for the work-life balance (cf. Moser 2007, p. 247 f.).

Apart from parent chats and work e-mails, the digitally increased frequency of exchange has an influence on informational self-determination. This is realized in the control of social roles and information flow. Together with other freedoms, this self-determination enables a social game with self-knowledge. It allows claims to power to be reflected or questioned, to form the image of a self internally and externally, and thus to take a place in the social structure. It belongs to the autonomous, free, mature and enlightened human being.

---

[4] In terms of the communication options described.

If the individual has a right to determine which information he or she releases to which addressees, he or she must know at all times in which role and in which context determining this role he or she is currently located. Informational self-determination also depends on transparency about who is in possession of what personal information at what point in time. Only with this knowledge can the shaping of one's image be attempted—a poker face with an open hand becomes a farce. In digital, communicative infrastructures, however, this knowledge of role, context and complicity is not accessible to the originator of the personal information:

- On the one hand, because in the latent mix of different roles, between dinner with the girlfriend, the quick email check on the smartphone and planning for the guys' night out on Facebook, a multiple role perception arises. The how and where of multiple simultaneous roles leads to challenging role management. The failure of this organization leads, for example, to amusing but embarrassing incidents for the sender when a romantic message to his girlfriend accidentally ends up in the message distribution list at work.
- On the other hand, the technical complexity makes informational self-determination more difficult. The use of corresponding infrastructures is usually done without understanding or even knowing the technical background.[5] The app needs access to all contacts, photos and the microphone, so this access is granted. The use of personal data, the complicity of the secret private, thus, becomes de facto intransparent. It is also not improved by the corresponding information in terms and conditions, the acceptance of which is a prerequisite for use, but which are almost reflexively clicked away. All in or not at all.

It has long been unclear on which platforms which personal data is processed, given the multitude of offers used. The information about the individual is correspondingly situational and incomplete, a mere glimpse of the individual. Just as the photo of a man with a naked torso in Majorca, when his name is searched on Google, are fragmentary images of a social life, which nevertheless create an image for third parties. Intentionally, as in the case of the Facebook profiles that only show good humour, or unintentionally, as in the case of the Mallorca photo. It is possible that the analogue counterpart does

---

[5] This is true, one might argue, of most technical objects: should one not drive a car, then, because one does not understand how an engine works? The difference lies in the respective role of the secret private: by driving excessively, one gives away no more than one's presence on the road. Excessive use of digital infrastructures, on the other hand, creates a nuanced image of the self without the user noticing and without the user having control over where and how this information is used.

not even know of the existence of its digital image, in any case it is unaware of its complete form. It is not only the voluntary entries in the social network profile. The digital self is nourished by search terms, recent online purchases, media consumption, access times and locations. The picture is incomplete, containing only those partial aspects that are exposed to digital access. And yet an image of the user is created from this collection of information (cf. e.g. Keen 2015, p. 8).

Sometimes the fragmentary, incomplete digital self comes into contact with its analog counterpart, the fragmentary digital self shapes the perception of its analog counterpart to some extent. For example, when it directs the flow of information along automatically assigned filters. A *filter bubble* that, based on analyzed user behavior, only presents the information that seems 'relevant'. The rest is blanked out, creating "a kind of digital thought prison" (Helbing et al. 2016, p. 57) that defines the analogue user along with their calculated, fragmented digital self. The user is defined by his data—not vice versa:

> Technologies actively shape our notion of the self; they even define how and what we think about it. They shape the contours of what we believe to be negotiable and nonnegotiable; they define the structure and tempo of our self-experimentation (Morozov 2014, p. 395).

A digital self emerges that is nourished by personal data, but over which the analog self has no disposal—adaptation or deletion are beyond reach. The private becomes a predictable estimate based on past actions. The duality between thought and action is erased to some degree by algorithmic capture and evaluation. In this way, the digital infrastructure restricts freedom of choice.

Giving up informational self-determination in this way means subjecting oneself to heteronomy. One accepts a heteronomous way of life: one's social role and the dissemination of information about the self are regulated from the outside. The consequence is a loss of control over one's own external appearance and a life that is determined from the outside without secrets (cf. Haberer 2015). In other words: informational heteronomy.

4.3.2 New Social Spaces: Digital Intentionality and Self-Policing

Through digital information networks, togetherness is expanding. New roles overlap, emerge, making informational self-determination more difficult. Self-perception and the perception of others are also determined by technical infrastructure. This opens up the private sphere, formerly a separate, self-determined area, to a latent public. Collectively, participation in digital exchange creates new social spaces with their own rules, manners and reputation mechanisms—*netiquette*. Disregarding these rules can lead to expulsion from the community.

Moving in these spaces is not a matter of choice. 'You can leave it alone' no longer works in relation to the digital infrastructure and its use, but would have far-reaching consequences for social exchange: communicatively, informatively and participatively, one would be cut off, professionally and in one's leisure time, civically and privately restricted. The attempt to circumvent the use of personal data and thus the external role formation of the digital self also seems difficult. This can be seen symbolically in the notification about the setting of cookies: the user is merely informed that a cookie has been set and that his behaviour on the site is being monitored and stored. He can only confirm this message, only take note that the monitoring is happening. He cannot press 'No', he can only leave the site. The same logic applies to terms and conditions, novel-length documents, the use of which implies acceptance. You do not want to read the long legal litany, but you reflexively confirm that you have read *it—take it or leave it.* The display of cookie information or the acceptance of the terms and conditions are not so much choices rather an *illusion of control,* like a button to close the elevator door or buttons at traffic lights that give the good feeling that one can influence the environment according to one's needs. In reality, the buttons are not even connected (cf. Thompson 2004).

The new digital social spaces are inevitable. And they have a normative effect. Instead of shaping themselves according to users' specifications, they lead to behavioral adaptations. Two of these adaptations will concern us specifically in the following: *digital intentionality* and *self-policing*.

4.3.2.1 Digital Intentionality

The inherent publicness of digital social behavior and its reflection lead to a higher degree of intentionality compared to analog social behavior. Digitalization rationalizes the private.

At first, this rationalization seems to be a pragmatic necessity: the expanded communication possibilities and partners make it imperative to order the information and contacts along certain criteria and comparisons. Those who can constantly contact everyone must reduce complexity if they want to remain capable of functioning. The techno-algorithmic selection that organizes social exchanges and pre-empts social behavior in the digital space does just that. Intentionality is a reaction to the uncertainty that accompanies the expansion of social exchanges.

This premeditation becomes emblematic at the various stages of the search for a partner. If we look back to the pre-digital era of dating, we experience a spontaneous social selection at the village festival, dependent solely on personal appearance and situational responsiveness. Intentional here is only the attendance of the event and the basic openness to exchange. The selection is based on the personal, lifeworld environment. Complicity in the process exists only as long as it is tolerated and until the chosen setting is left again. One digitization step further, in speed dating, those present are selected just as randomly, but getting to know each other with the intention of setting up a partnership is forced in a rationally planned manner: the registration for the event takes place with a fixed, calculated intention. The chances of finding a partner are higher if the persons willing to date are brought together in a planned manner. Thus, chance does not have a role in getting to know each other. The paid participation in the event, the temporal structure and rotation rationalize getting to know each other. Contingency decreases. This decline in contingency is even more pronounced when using an online partner exchange. Here, the criteria according to which the counterpart is selected are multiplied. Before the first written and then possibly personal exchange takes place, people are screened out. Creating one's profile and reflecting on the requirements of the person one is looking for are planned processes, a writing down or visualization of desires. A person is formed according to his or her wishes. The technical possibilities should help to find this perfect match. An undertaking that, so the conviction goes, would not succeed in analogue form without the planned digital search. Supported by the miraculous 'matching algorithm', the digital infrastructure processes the most secret wishes for the partner and throws out profile names. The dream man is just a click away. No wandering and tangling, but mathematical precision (cf. Bridle 2014). The pragmatic intentionality of searching entrusts itself to calculation.

Just as in the search for a partner, the calculative-judgement also has an effect on the area of sharing. While religious rationality was still the decisive factor in the St. Martin's famous act of sharing—guiding charity and godliness—and in the first charitable organisations environmental protection or consumer criticism, sharing is being opened up to a broader target group with the inclusion of digital infrastructures. The previously ideological leitmotifs are replaced by economic considerations. Behaviour is oriented towards economic intent. Visible, for example, when sharing a place to sleep via Airbnb: no spontaneous invitation, but professionally shot photos to advertise a private bedroom. Even the brief conversation with the successfully recruited short-term tenants to greet them becomes an expected exchange to secure a good rating: *always helpful, has the best recommendations for dinner.* Movement in the digital space is aligned with intentionality, planning and wanting, whether it is finding a partner or renting out your bedroom.

In addition to the rationalized wanting comes the constant storage and retrieval of interactions. The conversation feels different when the cell phone camera is rolling. People pull themselves together: *self-policing* as the order of the day.

4.3.2.2 Self-Policing Instead of the Right to Be Forgotten

Self-policing refers to self-imposed self-censorship due to a diffuse fear of repression (cf. e.g. Georgi 2016). The traceability and the eternal memory of the digital infrastructure make the participants of the new social spaces in the digital suitable for this form of restriction.

Analog social spaces are characterized by an interplay of remembering and forgetting, in which forgetting rather than remembering is what the human brain wants. Next to the limited capacity of memory, this seems to be the more merciful variant of togetherness (cf. Mayer-Schönberger 2011, p. 115). Forgetting, forgiving, and repressing simplify interactions with others. Missteps are calculated into the interpersonal and quickly forgotten again, at least if they are outweighed by positive experiences. An unobserved, oral communication has already disappeared in the moment of speaking, at most still reverberates in the memory of the participants, subjectively and blurred. This applies to the individual as well as social forgetting: it helps to get over the shortcomings of fellow human beings and the mistakes of the past (cf. Schneier 2015, p. 94).

In the digital age, forgetting no longer exists—the secret private is not only public but also stored and retrievable. Since the registration of the first personal mail account, since the first photo in the cloud, every single day that is somehow digitally accompanied can be reconstructed. The sender and the recipient each have a copy, and the operators of the infrastructure are either in direct possession of the stored communications or have their metadata at their disposal. Doing something about this seems hopeless: a 'right to be forgotten' in the digital space, as politically demanded (cf. Brössler 2015), is far from being efficiently implemented. Too much data is being collected too quickly about users in ever new ways, new systematics of identification are giving way to the old categories of persons that still prevail in politics. If a company is no longer allowed to work with a real name, it simply resorts to a unique user ID.

4 The Private Sphere Changes: A Consequence of Digitalization

Through its establishment, digital memory becomes part of the social control that is fundamentally inherent in humans. This control functions at a very low threshold: for example, a schematic representation of a face with three correspondingly arranged dots is sufficient to provoke a change in behaviour (cf. Norenzayan 2013, p. 21). The depicted eye in trains without a controller fulfils the same function. The effectiveness of these cues is demonstrated not least by their rejection, as the following illustration clearly comments with the calligraphic inclusion of the watchful eye (Fig. 4.3).

In the digital space, such an awareness of surveillance—and a critical reaction to it—also exists. It is even amplified due to the knowledge of the scope of digital memory. Through the spying on digital communication and the access possibilities to written and audiovisual communication as well as further sensor technology, surveillance is moving extremely close to the individual. It becomes virtually seamless, simultaneously possible on a massive scale

Fig. 4.3 Rejection of surveillance in public spaces. Own photograph clz, 01.06.2016, Allschwil BL

and impersonal: algorithms and automated keyword searches assess the likelihood of being involved in certain criminal activities. Human resources are drastically reduced. Privacy is combed through by a machine and processed for a potentially more detailed investigation.

This awareness leads to self-monitoring, *self-policing*, and avoidance strategies. Even those who do not engage in any illegal activity adapt their behaviour to the digital space, stick a black strip over their webcam, better not make jokes about Islamic terror in online forums and better not express their concerns about the new boss in an email. It is all being read (cf. e.g. Georgi 2016). Via smaller and larger scissors in the head, the awareness materializes that social behavior in the digital space can be monitored, interactions can be stored, usage behavior can be reconstructed. The constant recording makes it harder to say what one thinks, changes the character of exchange. Every statement and action is potentially relevant, perhaps even significant. Those who must always fear all possible consequences of their statements hold back. The dampening effect of suddenly having a camera or microphone pointed at the speaker in the middle of a conversation.

Self-censorship accepts the reproducibility of one's actions and statements by third parties as a fact and assumes latent surveillance as the guardrails of one's actions. What was initially conceived by Bentham in the form of the panopticon only as an architectural solution for schools, prisons and factories, now sits in the minds of the users of digital spaces and determines their social exchange.

References

Andrejevic, Mark (2016): Theorizing Drones and Droning Theory. In: Ders. (Hg.): *Drones and Unmanned Aerial Systems*. Heidelberg: Springer, S. 21–43.

Bamford, James (2012): The NSA Is Building the Country's Biggest Spy Center (Watch What You Say). *WIRED*. Abrufbar unter: https://www.wired.com/2012/03/ff_nsadatacenter/ *(letzter Zugriff: 03.06.2016)*.

Barrett, David (2016): Google to deliver wrong ‚top' search results to would-be jihadis. *The Telegraph*, Technology, February 2.

Becker, Gary S. (1991): *A Treatise on the Family*. Cambridge: Harvard Univ Press.

Bloom, Peter (2015): Authoritarian capitalism in modern times: when economic discipline really means political disciplining. *OpenDemocracy.net*. Abrufbar unter: https://www.opendemocracy.net/can-europe-make-it/peter-bloom/authoritarian-capitalism-in-modern-times-when-economic-discipline-rea *(letzter Zugriff: 20.04.2015)*.

Botz-Bornstein, Thorsten (2015): *Virtual reality: the last human narrative?* Leiden: Brill Rodopi.

Bridle, James (2014): The algorithm method: how internet dating became everyone's route to a perfect love match. *The Guardian*, Valentine's Day, February 9.

Brössler, Daniel (2015): Recht auf Vergessen: Schutz vor Google und facebook. *Süddeutsche Zeitung*, Digital, June 15.

CCC (2016): Satzung des CCC e.V. *CCC e.V.* Abrufbar unter: https://www.ccc.de/satzung *(letzter Zugriff: 08.06.2016).*

Chao, Loretta (2011): Cisco Poised to Help China Keep an Eye on Its Citizens. *The Wall Street Journal*, International, July 5.

Crouch, Colin (2000): *Coping with Post-Democracy.* London: Fabian Society.

Crueger, Jens (2013): Privatheit und Öffentlichkeit im digitalen Raum: Konflikt um die Reichweite sozialer Normen. *Aus Politik und Zeitgeschichte* 15–16 S. 20–24.

Dahlmann, Wolfgang (2015): Polizei: Keine Kinderfotos auf facebook posten. *c't Fotografie.* Abrufbar unter: http://www.heise.de/foto/meldung/Polizei-Keine-Kinderfotos-auf-facebook-posten-2849133.html *(letzter Zugriff: 09.06.2016).*

Daley, Jason; Piore, Adam; Lerner, Preston; et al. (2011): How to Fix Our Most Vexing Problems, From Mosquitoes to Potholes to Missing Corpses. *Discover Magazine* (October), S. 36–39.

Debord, Guy (1996 [1967]): *Die Gesellschaft des Spektakels.* Berlin: Edition Tiamat.

Engels, Friedrich (1956 [1877]): Herrn Eugen Dührings Umwälzung der Wissenschaft. In: Institut für Marxismus-Leninismus beim ZK der SED (Hg.): Berlin: Dietz-Verlag, S. 5–304.

Engquist, Erik (2015): Judge rules on taxi industry lawsuit: Compete with Uber or die. *Crain's New York Business.* Abrufbar unter: http://www.crainsnewyork.com/article/20150909/BLOGS04/150909863?template=print *(letzter Zugriff: 12.09.2015).*

Exenberger, Andreas (1997): Die Soziale Marktwirtschaft nach Alfred Müller-Armack. *Working Paper Institut für Wirtschaftstheorie, Wirtschaftspolitik und Wirtschaftsgeschichte Universität Innsbruck* 97 (1), S. 1–19.

Franck, Georg (2003): Mentaler Kapitalismus. In: Liessmann, Konrad Paul (Hg.): *Die Kanäle der Macht. Herrschaft und Freiheit im Medienzeitalter.* Wien: Zsolnay, S. 36–60.

Garland, David (2001): *The Culture of Control: Crime and Social Order in late Modernity.* Chicago: The University of Chicago Press.

Georgi, Oliver (2016): Staatliche Überwachung führt zu Selbstzensur im Netz. *Studie: Staatliche Überwachung führt zu Selbstzensur im Netz – Ausland – FAZ. pdf,* March 29.

Gonzalez, Juan (2012): „We Don't Live in a Free Country": Jacob Appelbaum on Being Target of Widespread Gov't Surveillance. *Democracy Now.* Abrufbar unter: http://www.democracynow.org/2012/4/20/we_do_not_live_in_a *(letzter Zugriff: 03.06.2016).*

Grassegger, Hannes (2014a): *Das Kapital bin ich. Schluss mit der digitalen Leibeigenschaft!.* Zürich: kein&aber.

Grassegger, Hannes (2014b): Jeder hat seinen Preis. *Zeit*, Konsum, October 27.

Haberer, Johanna (2015): Wahrheiten und Lügen – zur informationellen Selbstbestimmung. *Angewandte Ethik – BR Alpha CAMPUS Auditorium*. 8 Juni 2015.

Habermas, Jürgen (1990 [1962]): *Strukturwandel der Öffentlichkeit: Untersuchungen zu einer Kategorie der bürgerlichen Gesellschaft*. Frankfurt am Main: Suhrkamp.

Habermas, Jürgen (1998): *Die postnationale Konstellation. Politische Essays*. Frankfurt a. M.: Suhrkamp.

Hansen, Hans Krause; Flyverbom, M. (2014): The politics of transparency and the calibration of knowledge in the digital age. *Organization* S. 1–18.

Heath, Joseph; Potter, Andrew (2005): *The Rebel Sell. Why the Culture can't be jammed*. Chichester: Capstone.

Helbing, Dirk (2015): *The Automation of Society is Next How to Survive the Digital Revolution*. Zürich: ResearchGate.

Helbing, Dirk; Pournaras, Evangelos (2015): Build Digital Democracy. *Nature* 527 S. 33–34.

Helbing, Dirk; Frey, Bruno S.; Gigerenzer, Gerd (2016): Digitale Demokratie statt Datendiktatur. *Spektrum der Wissenschaft* (1), S. 50–58.

Heumann, Milton; Cassak, Lance; Kang, Esther; et al. (2016): Privacy and Surveillance: Public Attitudes on Cameras on the Street, in the Home and in the Workplace. *Rutgers JL & Pub. Pol'y* 14 S. 37–83.

Hill, Kashmir (2015a): Cops are asking Ancestry.com and 23andMe for their customers' DNA. *fusion*. Abrufbar unter: http://fusion.net/story/215204/law-enforcement-agencies-are-asking-ancestry-com-and-23andme-for-their-customers-dna/ *(letzter Zugriff: 20.10.2015)*.

Hill, Kashmir (2015b): facebook will now be able to show you ads based on the porn you watch. *fusion*. Abrufbar unter: http://fusion.net/story/199975/facebook-tracking-like-buttons-for-ads/ *(letzter Zugriff: 20.10.2015)*.

Hopkins, Nick; Borger, Julian (2013): Exclusive: NSA pays £100m in secret funding for GCHQ. *The Guardian*, UK, August 1.

Jarvis, Jeff (2011): *Public parts: How sharing in the digital age improves the way we work and live*. New York: Simon and Schuster.

Keen, Andrew (2015): *The Internet is not the answer*. New York: Atlantic Books Ltd.

Kire (2015): Überwachung. *Digitale Gesellschaft*. Abrufbar unter: https://www.digitale-gesellschaft.ch/ueberwachung/ *(letzter Zugriff: 01.12.2015)*.

Langfitt, Frank (2013): In China, Beware: A Camera May Be Watching You. *NPR*. Abrufbar unter: http://www.npr.org/2013/01/29/170469038/in-china-beware-a-camera-may-be-watching-you *(letzter Zugriff: 05.05.2015)*.

Lanier, Jaron (2014): *Who owns the future?* New York: Simon and Schuster.

Lapowski, Issie (2015): Google's Next Moonshot: Lining City Streets With Wi-Fi Hubs. *Wired*. Abrufbar unter: http://www.wired.com/2015/06/google-next-moonshot-wifi-hubs-sidewalk-labs/ *(letzter Zugriff: 07.06.2016)*.

Levison, Ladar (2014): Secrets, lies and Snowden's email: why I was forced to shut down Lavabit. *The Guardian*, Opinion, May 20.

Lobo, Sascha (2014): Auf dem Weg in die Dumpinghölle. *SPIEGEL ONLINE*. Abrufbar unter: http://www.spiegel.de/netzwelt/netzpolitik/sascha-lobo-sharing-economy-wie-bei-uber-ist-plattform-kapitalismus-a-989584.html *(letzter Zugriff: 20.10.2015)*.

Marx, Karl (1956 [1859]): Der Fetischcharakter der Ware und sein Geheimnis. In: Marx, Karl; Engels, Friedrich (Hg.): *Werke. Band 23. Herausgegeben vom Institut für Marxismus-Leninismus beim ZK der SED*. Berlin: Dietz-Verlag, S. 85–99.

Marx, Karl; Engels, Friedrich (1962 [1848]): Das Manifest der kommunistischen Partei. In: Marx, Karl; Engels, Friedrich (Hg.): *Werke. Band 18. Herausgegeben vom Institut für Marxismus-Leninismus beim ZK der SED*. Berlin: Dietz-Verlag, S. 352–493.

Mayer-Schönberger, Viktor (2011): *Delete: The Virtue of Forgetting in the Digital Age*. Princeton/Oxford: Princeton University Press.

McCloskey, Deirdre (2006): *The Bourgeois Virtues: Ethics for an Age of Commerce*. Chicago: University of Chicago Press.

Morozov, Evgeny (2014): *To save everything, click here: The folly of technological solutionism*. New York: PublicAffairs.

Moser, Klaus (2007): *Lehrbuch Wirtschaftspsychologie*. Heidelberg: Springer.

NDR (2015): Schöne neue Welt – der Preis des Teilens. *Transkript zur Sendung Panorama, 08.01.2015*. Abrufbar unter: http://daserste.ndr.de/panorama/archiv/2015/panorama5378.pdf *(letzter Zugriff: 17.09.2015)*.

Norenzayan, Ara (2013): *Big gods: How Religion Transformed Cooperation and Conflict*. Princeton: Princeton University Press.

Reuters (2016): Obama wirbt für Zugriff auf Handys in Ausnahmefällen. *Reuters*. Abrufbar unter: http://de.reuters.com/article/usa-apple-obama-idDEKCN0WE0NO *(letzter Zugriff: 13.03.2016)*.

Robbins, Lionel (1945): *An Essay on the Nature and Significance of Economic Science*. 2. Aufl. London: Macmillan.

Sandberg, Sheryl (2011): Sharing to the power of 2012. *The Economist*, United States, November 17, The World In 2012.

Savage, Charlie (2016): Obama Administration Set to Expand Sharing of Data That N.S.A. Intercepts. *The New York Times*, Politics, February 26.

Schneider, Manfred (2015): Ende des Gesellschaftsvertrages, Aufstieg der Überwachungskultur. *NZZ-Podium vom 24. September 2015 «Überwachungskultur»*. Abrufbar unter: http://podium.nzz.ch/event/uberwachungskultur/ *(letzter Zugriff: 29.09.2015)*.

Schneier, Bruce (2015): *Data and Goliath: The hidden battles to collect your data and control your world*. New York: WW Norton & Company.

Schulz, Thomas (2015): Wetten auf die Zukunft. *DER SPIEGEL* (34), S. 18–19.

Seele, Peter; Zapf, Lucas (2014): 'The Markets Have Decided': Markets as (Perceived) Deity and Ethical Implications of Delegated Responsibility. *Journal of Religion and Business Ethics* 3 (17), 1–21.

Sennett, Richard (2010): *Der flexible Mensch. Die Kultur des neuen Kapitalismus*. 7. Aufl. Berlin: BTV.

Singer, Natasha (2014): Listen to Pandora, and it listens back. *New York Times*, Technophoria, Technology.

Singer, Natasha; Isaac, Mike (2015): Uber Data Collection Changes Should Be Barred, Privacy Group Urges. *The New York Times*, June 23, New York.

Softpedia.com (2015): AVG Proudly Announces It Will Sell Your Browsing History to Online Advertisers. *Softpedia.com.* Abrufbar unter: http://news.softpedia.com/news/avg-proudly-announces-it-will-sell-your-browsing-history-to-online-advertisers-492146.shtml *(letzter Zugriff: 21.09.2015).*

statista (2016): Anzahl der verschickten SMS- und WhatsApp-Nachrichten in Deutschland von 1999 bis 2014 und Prognose für 2015 (in Millionen pro Tag). *Statistik-Portal.* Abrufbar unter: http://de.statista.com/statistik/daten/studie/3624/umfrage/entwicklung-der-anzahl-gesendeter-sms-mms-nachrichten-seit-1999/ *(letzter Zugriff: 25.05.2016).*

Stokowski, Margarete (2016): Nimm die Hand aus der Hose, wenn ich mit dir rede. *SPIEGEL ONLINE – Kultur.* Abrufbar unter: http://www.spiegel.de/kultur/gesellschaft/hass-im-netz-brief-an-den-unbekannten-hater-a-1090934-druck.html *(letzter Zugriff: 26.05.2016).*

Thompson, Suzanne C. (2004): Illusions of control. In: Pohl, Rüdiger F. (Hg.): *Cognitive Illusions A Handbook on Fallacies and Biases in Thinking, Judgement and Memory.* East Sussex: Psychology Press, S. 115–126.

Uchatius, Wolfgang (2013): Die Macht meines Kreuzes: Soll ich wählen oder shoppen? *ZEIT*, Politik, September 19, 39.

Vincent, James (2016): The revolving door between Google and the White House continues to spin. *The Verge.* Abrufbar unter: http://www.theverge.com/2016/1/14/10766864/google-hires-white-house-personnel-caroline-atkinson*(letzter Zugriff: 31.05.2016).*

Weiss, Allan B. (2015): *Memorandum Supreme Court Queens County: Uber and New York Taxis.* Queens: Supreme Court.

Werner, Jürgen (2014): Markt. In: Werner, Jürgen: *Tagesrationen – Ein Alphabet des Lebens.* Frankfurt a. M.: tertium datur, S. 169–170.

Wollenhöfer, Thorsten (2014): Eltern- und Lehrer-Kommunikation – Mobbing 3.0? *SOCIALMEDIALERNEN.com.* Abrufbar unter: https://www.socialmedialernen.com/schulen/eltern-und-lehrer-kommunikation-mobbing/ *(letzter Zugriff: 24.05.2016).*

World Under Watch (2012): *Google is Watching you.*

Yadron, Danny; Wong, Julia Carrie (2016): Silicon Valley appears open to helping US spy agencies after terrorism summit. *The Guardian*, tech, January 8.

Zizek, Slavoj (2009): *First as Tragedy, then as Farce.* London, New York: Verso.

Zuboff, Shoshana (2016): Wie wir Googles Sklaven wurden. *faz*, Feuilleton, March 5.

5

Summary: Thoughts in a Digital World: Free, but no Longer Secret

'Thoughts are free'—this lyrical manifestation of inner autonomy first spread in the eighteenth century as a poem, later as a song. The message: despite autocratic systemic constraints, state persecution and repression, there is at least the inner life, the thoughts that no one can control—'I think what I want […] but everything in the still' […]' What clearly emerges is the need to control one's own desires and wants, in a self-determined manner and without outside influence. What is also clear, however, is that all this must remain within. The escape from the state's inquisitiveness into an inner self-knowledge.

Today, thoughts are probably still free, but the lyrics would have to be updated: thoughts are no longer secret. 'No man can know them' is no longer true for the digital age and the structural change of the private sphere. State encroachments on the citizen are now bound by law, but thanks to digital infrastructures the outward curiosity gets close to the inner sphere of the citizen, right down to his thoughts: what he is looking for, desires, purchases, hears, sees, where, when, for how long. Thoughts are still free, but the state knows them. And not only the state: desires conceived in silence are subject to corporate consideration and evaluation. No more 'nocturnal shadows', now thoughts are stored on the back of the cloud, where you cannot even look yourself. Thoughts are free, but they are no longer secret. The knowledge of their visibility, the lifting of their secrecy and their systematic use have an effect on the understanding of the private—and on the thoughts themselves.

The translation of this chapter was done with the help of artificial intelligence (machine translation by the service DeepL.com). A subsequent human revision was done primarily in terms of content.

This *theory of privacy without secrecy in the digital age is* about this change. The old concept of privacy admitted the possibility of a secret in the private sphere. It was based on the demarcation from the public sphere, on informational self-determination and thus on *self-knowledge*. This understanding no longer fits the digital age, in which personal information and communications are stored, globally exchanged and used in a flourishing trade. A *digital omniscience* prevails.

The abolition of secrecy in private is a profound social change that entails a structural transformation of the private sphere. Thoughts are still free, but there are confidants. Consequently, the notion of privacy needs to be updated. We have worked out this update and the theoretical considerations in which it is embedded in a three-step process:

- Part A: Typology of the secret private, therein the definitions of 1. *transcendent, analogical omniscience*; 2. *immanent, analogical 'intrinsic knowledge'*; and 3. *immanent, digital omniscience*.
- Part B: Description of symptoms and
- Part C: Development of a theory of the *structural change of the private*.

In the following, we summarize the results of these steps. The result is a description of the individual consequences of the described update of the concept of privacy.

5.1 Typology of the Secret Private

Our typology of Part A places the relationship between secrecy and privacy in a threefold, chronologically organized order. The two emerging transitions are

- the change from transcendence to immanence (divine omniscience vs. worldly inquisitiveness and omniscience), and
- the change from analogue to digital.

In considering these transitions, *secrecy*—as the absence of knowledge of or control over personal information—stands as the central feature of the private sphere. We present its transformation up to its abolition at the individual level (Sect. 2.2). The Barbie doll, which thanks to artificial intelligence talks to small children and simultaneously records what they say and transmits it to their parents or third parties, is an ideal example of this *private sphere without secrecy*. Mass society and individualisation are described as the social preconditions for this development (Sect. 2.3).

5 Summary: Thoughts in a Digital World: Free, but no Longer Secret

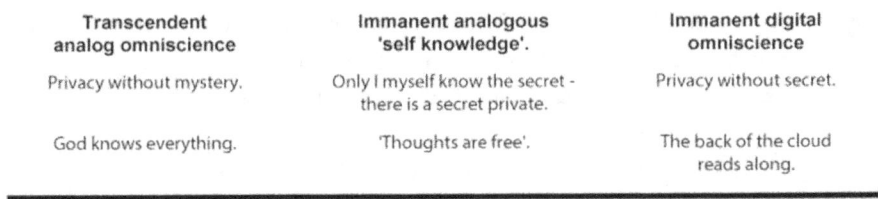

Fig. 5.1 Summary of the three types of secret privacy

To make the structural change of the private visible, we propose a typology of how privacy and secrecy relate to each other. Within the three types of the secret private we are in type 3 in the digital age (Fig. 5.1).

Privacy in the digital age is no longer protected by a secret. It is to be understood in the context of an immanent digital omniscience and puts pressure on the secret private from two sides:

1. Through mass digital information collection;
2. Through the individual traceability of the collected information.

The collection and storage of information are, firstly, the precondition for the abolition of the secret private: with the accumulated knowledge it is possible to reconstruct complex realities of life, and to a certain extent even to predict them, without the originator of the information knowing about the storage and processing, being involved in it, or being able to dispose of it. Secondly, this becomes virulent with the secret private because of the individual traceability of this information. While the (anonymous) collection of information—like any statistical analysis—enables statements to be made about group behaviour, in digital infrastructures this information can be traced back to individual persons. Thus, the collected knowledge becomes individually relevant information, while the originator of the information does not know to whom and when which knowledge about him is available. This results in informational heteronomy with regard to one's data.

5.2 Symptoms and Theory of Structural Change

The change of the private along the abolition of the secret can be traced in everyday life—in part B we show corresponding symptoms (Sects. 3.1, 3.2, 3.3, 3.4, 3.5, 3.6, 3.7, 3.8, 3.9 and 3.10). One example is taxi driving: in the analogue, one boards a taxi anonymously, tells the driver his destination and finally pays anonymously by cash. With the introduction of taxi apps, anonymity is reduced by electronic customer profiles. And finally, with the ride service provider Uber, all personal information, including payment details, route information and personal evaluation of driver and passenger, is digitally recorded and stored.

The typology of the secret private serves as a thread along which the symptoms of the structural change of the private are examined. Subsequently, the change is abstracted along the three areas of economy, politics and social affairs and summarized in the theory of a structural change of the private sphere (Sects. 4.1, 4.2 and 4.3). The leitmotif here is that, with digitalisation, the economy, politics and the social sphere are moving closer than before to the individual and his or her secret private sphere. The private sphere is subject to independent change in the various areas:

- *Economic structural change* is central to the overall structural change of the private sphere. It is the area in which the greatest changes in the private sphere are taking place in terms of the number of individuals affected and the extent of data used. With great entrepreneurial commitment, more and more new products are being developed that make significant use of personal information—virtually networked, the secret private sphere is becoming an individualized business case. By approaching the secret of the individual, the economy changes, uses, and shapes the secret private.
- The course of the *political structural change of* the private sphere is set in accordance with the economic paving of the way. Personal data are collected and evaluated with political motivation. Private information of the citizen becomes political information that can be used precisely in the election campaign. This creates a contradiction to the traditional understanding of privacy, in which *the private sphere* was explicitly understood as a counterpart to the *public sphere*. Likewise, ethical-legal contradictions arise with regard to attempts to prohibit economic actors from those practices with which state organs realize access to the secret private sphere. Overall, state organs come closer to the citizen, thereby using and shaping the secret private of the individual.

We observe a parallel development between economics and politics. The different objectives are each pursued through the collection, storage and evaluation of individually locatable private information, whereby economics and politics are closely intertwined. Due to the largely private-sector organisation of digital infrastructures, on which politics is dependent, this is associated with an increase in the power of economic actors.

Domain	Structural change
Economy	Economic use of private data and shaping of the secret private through consumption decisions. Expansion of economic activities to all areas of life
Politics	Entirety of personal information is exposed to political use. Contradictionto the formal legal concept of privacy.
Social	Creation of new social spaces with total digital behavior and behavioral adaptation.

Fig. 5.2 Structural change of the private sphere on the economic, political and social levels

- The economic and political structural change of the private sphere has effects as a *social structural change:* first, digital infrastructures create new social spaces. Opportunities for exchange multiply and become cheaper, subsequently used en masse and extensively. At the same time, the awareness of full digital documentation—within these spaces but also in the surrounding, only indirectly used digital infrastructure—prevails. This leads to latent reflection and consequently to the adaptation of one's own behavior. In short: to self-policing and intentionality in social interaction.

The following illustration summarizes the results of the theory section along the three areas investigated (Fig. 5.2).

The transformation of the secret private in the economic, political and social spheres is not only of systemic interest. As we continue to show in our theory, it also has serious consequences at the level of the individual.

5.3 Digital Formation of the Secret Private

The more intensively and comprehensively digital infrastructures are used, the closer and more precisely companies and the state move to the individual. As a consequence of this development, individual access to the secret private sphere is being shaped along digitally induced changes:

1. *Awareness of the secret private sphere:* Awareness of latent joint knowledge changes the perception of one's own privacy. Even if digital secrets are rarely actually communicated to third parties, the potential disclosure of

the information alone has a dampening effect: the knowledge of shared knowledge cancels out the secret of the secret private.
2. *Outsourcing of the secret private:* The data of the secret private and their combination with the data of others allow knowledge and conclusions of which the originator himself has no idea.

   The awareness and outsourcing of the secret private reinforce the reflection on the latent digital omniscience and eventually leads to a behavioral adaptation:
3. *Abandonment of the secret private:* Those who are aware of constant surveillance and know less about themselves than the surveillant can no longer realize their own notion of the secret private. Self-censoring and *self-policing ensue* and prevent thoughts that would be unpleasant and actions that are anticipated as inappropriate.

Back to the folk song, *Thoughts are free (Die Gedanken sind frei)*. Formally, still correct. But thoughts are no longer secret. If the song said: *No man can know them,* this line no longer applies today. Rather, thoughts are the subject of economic and political use, are processed and reflected back to their originator under new auspices. Informational self-determination is not envisaged in this process; on the contrary, one's own information is used manipulatively, as is done in marketing or *nudging* (cf. Helbing 2015).

References

Breuer, Hans (1920): *Der Zupfgeigenhansl.* Leipzig: Hofmeister.
Helbing, Dirk (2015): „Big Nudging" – zur Problemlösung wenig geeignet. *Spektrum der Wissenschaft* (11), S. 15–17.

6

Conclusion: Our Secrets Behind the Cloud

6.1 The Updated Concept of the Secret

As stated in the previous chapter, in the analogue age: *Thoughts are free*—no man can know, hunt or hunt them down. They are secret and not subject to control by powerful people or rulers, nor divine omniscience. Free thoughts are an expression of a secret private, of the knowledge of one's information and the control of what is shared with whom.

In the digital age, this understanding no longer seems current, it almost seems romantic and outdated. The secret private sphere has been abolished. What is seen, heard or communicated in the digital world is no longer subject to individual memory. The search engine recognizes everything from religious radicalization to illnesses that move the searcher in his innermost being. Elsewhere, the abolition of the secret private sphere is turning into entertainment: in reality shows, you can see fellow human beings sleeping, watch people on the other side of the world eating dinner, zap into the educational failures of precarious families or the courtship of wealthy city dwellers. When this entertainment is consumed on the smart TV, the voyeuristic viewer becomes an observer himself, tracking consumption, registering to dwell time and attention. Surrounded by one's own four walls and one's own devices, private activity only takes place with others 'behind the cloud' knowing.

The 'secret' in the digital is a fluid construction. The secret perhaps in the sense of password-protected-secret, protected from access by strangers if you lose your mobile phone. A mirrored screen that protects from prying eyes in

---

The translation of this chapter was done with the help of artificial intelligence (machine translation by the service DeepL.com). A subsequent human revision was done primarily in terms of content.

the suburban train. But never secret from digital access, from the back of the cloud. The secret detaches itself from a substantial and insurmountable actual secret. It detaches itself from those free thoughts that are the expression of complete self-knowledge and informational self-determination. The secret becomes a contextual, soft attribute. 'Secret' in the new sense is the privacy package on the new car with the tinted windows. But the onboard computer is constantly connected to the manufacturer's servers and stores location, speed, fuel consumption, entertainment program and even eye movements in case of impending fatigue.

Hence, our observation of structural change: the private is no longer secret in the digital. The preceding remarks call this actualization to mind. The following sections condense the consequences of this update.

6.2 The Secret Private in the Realm of Machines

Encroachments on the secret private have so far been associated with indiscretions or human actors: the insecure husband peeping into the wife's diary. The private detective or the secret agent with the binoculars in the opposite window. With the digitization of large parts of communication, evaluation and surveillance processes are increasingly automated: "[The] subtraction of the human element from the decision loop follows naturally from the automation of data processing" (Andrejevic 2016, p. 23). Access to aggregated personal data is initially automated and without human contact. With the amount of data to be processed, human analysis of the information is not even possible. Now, it is machines that have the knowledge of the individual secret private, stored somewhere in the distance on a server. This raises the question of non human intrusion into the secret private. What do the collecting machines know about us?

For the machine evaluation, the content of the private secrets searched is irrelevant. It is not about the content, but about the spirit, not about the knowledge of the conversation itself, but the intention. Computational models extract the desired information from the mass of data by evaluating signal words or the sequence of search terms. It computes what is between the lines, in tens of thousands of cases simultaneously. The calculation is complemented by knowledge of the associated metadata: location, medium, duration, destination, complements the machine evaluation of the data and its assessment.

The machine intrusion into the secret private sphere is put into perspective by the users of the collected data—the masters of the machines. Thus, Google responds to critical voices that object to the evaluation of emails for the personalization of advertising: "The process by which ads are shown in Gmail is fully automated. Nobody reads your emails in order to show you ads" (google 2016). At Google, nobody reads the emails. It is just software that extracts keywords and then assigns appropriate advertising messages to accounts. Google considers the machine intrusion into the secret private of its users to be unproblematic.

On closer inspection, however, the relativization is not very plausible. Rather, the statement illustrates that the machine-generated and mass-collected results allow highly individualized conclusions to be drawn: trends can be read from the aggregated individual data, which can then be compared with the individually assignable data set. In this way, an individual shopping recommendation for the next supermarket visit is derived from the initially anonymous, machine-generated data. Alternatively, and in the case of other calculations, the police will be at the door 10 min after the wrong search word combination.

Relevance to life and death this mechanism receives the so-called *signature strike*. This refers to the strike against a person against whom there is no personal evidence, but who is most likely guilty due to the circumstances. A data-driven profile is sufficient to determine guilt; personal identification is not necessary. The person disappears behind a signature and the drone attack does the rest (cf. Rohde 2015). Machine data analysis and the reflection back on individual behavior as a death sentence—data collection more precise than personal identification. The computer does not need a name to identify a guilty person. Not dystopian science fiction, but daily practice in the US fight against international terrorism.

Less sensationally but no less effectively, another area provides information about the nature of machine evaluations of the secret private. The starting point: discrimination by digital infrastructures, exemplified by the practice of *dynamic pricing*. Here, prices are adjusted on the basis of time of day, access medium and route to the online shop, i.e. by combining personal data with aggregated information, in order to achieve the highest possible willingness to pay on the part of the customer (cf. SWR 2015). Those who shop from their tablet in the evening pay more than those who shop from their PC in the morning. Search engine algorithms can also discriminate, as shown not least by lawsuits against Google's auto-complete mechanism: celebrities feel offended by the suggestions that appear after entering their name. It is unpleasant when terms, such as "cheating" or "liar" appear together with an individual name. Admittedly, the ad does not appear willfully malicious but merely anticipates most query combinations.[1] However, according to the critics' argumentation, this means that the algorithm anticipates the individual search decision and guides the unsuspecting surfer in a discriminatory manner.

If we abstract the examples considered, we can answer the question of whether machine-driven intrusion into the secret private sphere appears less problematic than a human intrusion, in summary, in the negative from two points of view:

1. The technical infrastructure and connected machine surveillance is an expression of a human-made system and thus subject to human goals. Even if the surveillance in the concrete case is represented by a non-human agent, such as an algorithm;
2. The fact that the automatically collected findings can be stored as well as retroactively retrieved and used turns any machine-collected information and its analysis into information potentially accessible to the human eye.

Accordingly, the machine evaluation of personal data cannot be separated argumentatively from a personal-human evaluation: the underlying purposes of surveillance remain the same. Also, the panoptic knowledge effect that leads to behavioral adaptations remains the same, regardless of whether the confidant of the secret is human or machine.

6.3 What to Do?

As a consequence of the abolition of the secret private in the digital sphere, we want to ask the question: what to do? What life-world consequences are to be drawn from the results of the present study?

In the face of a private without secrets in the digital age, we see three possible ways out of self-inflicted informational heteronomy: (1) creating an awareness of the structure of those infrastructures that led to actualized privacy and knowledge of the goals that lie behind these infrastructures. (2) consciously dealing with the infrastructures that attack the secret private. And (3) an emerging market that could fix it: *privacy* as a business case.

---

[1] This function holds interesting possibilities: If you want to know what people are currently thinking about, you only need to enter "I hate", "I am afraid of", "I definitely do not want", etc. and observe what the search engine adds (cf. Lange-Hausstein 2016).

6.3.1 Making People Aware of What They Have Made

Behind the use of the secret private in the digital sphere lies a multitude of man-made interests (cf. Hables Gray 2016). Consequently, it is important to understand the encroachments on the secret private not as an independent, unbridled and incomprehensible power, but as a respective means to an end—be it social, economic or political. The ends are the starting point for abolishing secrets. Changing these ends is the most powerful means to do so: those who reduce the digital connectivity of their social lives limit the access of social networks to their privacy. Those who do not willingly open their personal information to marketing are less likely to receive personalized advertising. Whoever favours a policy that places civil rights above the feeling of security restricts the access of state surveillance to one's private sphere.

There must be a reflection on the composition and malleability, because only in this way can a collective ethical framework for digital infrastructures and their new social spaces emerge. These rules must be capable of being shaped by users and providers alike (cf. Floridi 2014, p. 236).

This layout of digital structures, the linkage with the interests of their creators, and the need for an ethical framework also apply to further technological development. Machine extensions of humans, such as the idea of a techno-humanoid (cf. Allenby and Sarewitz 2011) or a *superhuman condition* in which humans and machines each merge functionally and in terms of consciousness (cf. Morozov 2014, p. 312 ff.), are subject to the same system. If artificial intelligence is able to develop independently, sooner or later the question arises as to when one should grant these made structures an independent status, or at what point the creation separates itself from its creator and must be perceived as autonomous. The structures considered here, however, are removed from this autonomy and thus autonomy of their own, and are subject to and attributable to clear purposes, goals, and thus human creators. Behind every algorithm is a programmer and behind this a company, a state or another interest.

6.3.2 Conscious Use of the Digital Infrastructure

The starting point for the conscious use of digital infrastructures is the understanding of the strength that the secret private has for personal development: "The goal of privacy is not to protect some stable self from erosion but to create boundaries where this self can emerge, mutate, and stabilize" (Morozov

2014, p. 398). The secret private creates strength through calm and concentration, which stabilizes the individual (cf. Lotter 2013, p. 26).

Necessary for a comprehensive awareness would be the knowledge of the functioning of all surrounding devices, which requires a high level of technical savvy. In addition, data collection often takes place unnoticed: whether the camera is recording or the app is accessing the location is not always apparent. Initially, therefore, it would be a matter of securing the ownership rights to the personal information, limiting its availability and, if necessary, profiting from its disclosure (cf. Pentland 2014). Currently, privacy in this form is an elitist affair of those who see through the structures, master them, or do not need them. Of techies and people who can afford to do without the digital. Mark Zuckerberg may claim that privacy is no longer a social norm. At the same time, however, he is buying up the four neighbouring houses around his mansion in Palo Alto to guarantee his privacy (cf. Schneier 2015, p. 92). In this context, the normal working man is left with fewer places of retreat where he can feel safe from encroachment on his secret private sphere.

The latent threat to privacy calls for an awareness of the role one is currently in. Given the multiplication of roles in the new social spaces, the intrusion of economic and political interests, and the overlaps that arise, this is not an easy task. Becoming role-aware requires some reflexive effort. Those who continuously disclose personal data and allow sensors to measure their immediate surroundings quickly lose track of the role they are occupying: as a customer, citizen or friend? Or as a private person? But awareness of the different roles, once accomplished, can be used to benefit privacy. And it can be done with the help of obfuscation. Tried and tested in intelligence work now in use for privacy in the digital space. A proven tool for "those at the wrong end of asymmetrical power relationships" (Brunton and Nissenbaum 2015, p. 22). This involves disseminating ambiguous, confusing or deliberately false images of oneself in different roles, thereby compromising surveillance and data collection. Multiple identities for different roles are crafted to one's liking. The image that emerges to the outside world remains inconclusive. The actual, secret private is protected.

6.3.3 Privacy as a Business Model

The reflection on the made nature of digital infrastructures and consequently the more conscious handling create a demand for products that respect the secret private in the digital, respect the boundaries between customers and companies. Privacy must already be anchored in the development and at all levels of the product, and all aspects that endanger privacy must be deactivated by default: *Privacy by Design* and *Privacy by Default* are the corresponding keywords.

The global corporation Apple also seems to see the growth potential of products of this kind. With its integrated offerings, the company is very close to its users: *wearables* like the Apple Watch, cross-platform synchronization of address book, e-mail and user data, the iPhone always at hand. A loss of trust in the security of the personal data handed over would be problematic. And so the company is trying to use the privacy issue to its advantage. The linchpin of Apple's strategy is encryption, arguably the most efficient means of protecting the secret private in the digital. However, the encryption of large manufacturers usually has a shortcoming: a so-called *backdoor*, through which the encryption can be bypassed for service or development purposes. A backdoor that is also used by government agencies to monitor the encrypted data of suspected criminals. The issue of encryption and *backdoor* was widely discussed in the case of an iPhone-using assassin who carried out an attack in the US. The FBI demanded that Apple decrypt the assassin's mobile phone. Apple refused, arguing that they did not want to provide access to the private lives of their customers—and had developed their products in such a way that this was not even possible. The US authorities did not believe the company, and heated discussions, charges against the company and political pressure followed. In the end, however, the FBI had to come to terms with Apple's position. Tim Cook, CEO of Apple, meanwhile published a detailed statement on the company's homepage:

> Customers expect Apple and other technology companies to do everything in our power to protect their personal information, and at Apple we are deeply committed to safeguarding their data. Compromising the security of our personal information can ultimately put our personal safety at risk. That is why encryption has become so important to all of us. For many years, we have used encryption to protect our customers' personal data because we believe it's the only way to keep their information safe. We have even put that data out of our own reach, because we believe the contents of your iPhone are none of our business (Cook 2016).

The trust of customers in Apple that their personal data will not be exploited or passed on is essential for the further use of the products—even more so than in the area of product design or execution itself (cf. Crane and Matten 2015). Privacy as a business case comes into play when there is a market for it.

6.4 In Conclusion

The digital infrastructure is a means for the most diverse purposes, economic, social or political: some want to earn money, others want to assert their interests, and some want both. In the process, the secret private is used, knowingly and unknowingly. This puts informational self-determination into perspective. And all of this is done largely of one's own free will: personal data is revealed without coercion and freely, processing accepted, in social networks, in mobile phone use, in online shopping. We are dealing with a self-inflicted immaturity, a consequence of hastily accepted terms and conditions and the unhesitating feeding of personal data into the cloud, to the back of which we no longer have access. Self-inflicted informational heteronomy. The historically short phase of self-knowledge, in which we were masters of our data, is passé. The back of the cloud sometimes knows us better than we know ourselves—and we currently have no access to this knowledge ourselves. Others, however, do—and that is the real point of criticism. Digital omniscience is therefore on its way to being established.

The secret private, it seems to us, is gone.

References

Allenby, Braden R; Sarewitz, Daniel (2011): *The Techno-Human Condition*. Cambridge: MIT Press.
Andrejevic, Mark (2016): Theorizing Drones and Droning Theory. In: Ders. (Hg.): *Drones and Unmanned Aerial Systems*. Heidelberg: Springer, S. 21–43.
Brunton, Finn, Nissenbaum, Helen (2015): *Obfuscation: A User's Guide for Privacy and Protest*. Cambridge: MIT Press.
Cook, Tim (2016): A Message to Our Customers. *apple.com*. Abrufbar unter: http://www.apple.com/customer-letter/ *(letzter Zugriff: 18.02.2016)*.
Crane, Andrew; Matten, Dirk (2015): Apple's big bet on consumer trust and privacy. *Crane and Matten blog*. Abrufbar unter: http://craneandmatten.blogspot.ch/2015/03/apples-big-bet-on-consumer-trust-and.html *(letzter Zugriff: 19.05.2015)*.
Floridi, Luciano (2014): *The fourth revolution: How the infosphere is reshaping human reality*. Oxford: Oxford University Press.
google (2016): How Gmail ads work. *Gmail Support*. Abrufbar unter: https://support.google.com/mail/answer/6603?hl=en *(letzter Zugriff: 07.01.2016)*.
Hables Gray, Chris (2016): Could Technology End Secrecy? *Secrecy and Society* 1 (1), S. 1–7.

Lange-Hausstein, Christian (2016): Lücke im Recht: Vom Algorithmus diskriminiert. *SPIEGEL ONLINE – Netzwelt.* Abrufbar unter: http://www.spiegel.de/netzwelt/web/digitale-diskriminierung-luecke-zwischen-algorithmus-und-mensch-a-1082219.html *(letzter Zugriff: 21.03.2016).*

Lotter, Wolf (2013): Die Ruhestörung. *Brand Eins* 08/13 (Schwerpunkt: Privat), S. 22–27.

Morozov, Evgeny (2014): *To save everything, click here: The folly of technological solutionism.* New York: PublicAffairs.

Pentland, Alex (2014): *Social physics: How good ideas spread-the lessons from a new science.* New York: Penguin.

Rohde, David (2015): Warren Weinstein and the Need to End ‚Signature' Drone Strikes. *The Atlantic*, International, 04/15.

Schneier, Bruce (2015): *Data and Goliath: The hidden battles to collect your data and control your world.* New York: WW Norton & Company.

SWR (2015): Online-Shopping Preistricks zu Lasten der Kunden. *SWR Fernsehen.* Abrufbar unter: http://www.swr.de/marktcheck/dynamic-pricing/-/id=100834/did=16117742/nid=100834/1y8yazo/index.html *(letzter Zugriff: 30.06.2015).*

7

Outlook: Digital Authenticity: Immersive Consumption Without Secrets

At the end of the year, in December, retrospectives are on the agenda. Not only television but also digital channels are full of them. A special service: social media puts together a personal year in review for its users. The algorithm generates the individual highlights of the last 12 months. The posts with the most likes, the pictures with the most comments are rated as the most important stations. The digital photo album also checks in with an automatic selection. The pictures that have been viewed or sent the most are presented and let the significant experiences pass in review. The digital omniscience knows the user and what has moved him this year. What should be remembered? And what was not so important. An authentic image of one's own life, digitally reformulated by personal data. The result is an immersive experience: an immersion in the virtual-economic environment. This economic finesse of reflecting the user's 'real' own life and experience and selling it as authentic is the subject of the following question. How do companies succeed in generating this authenticity? (cf. Zapf (2020) on this question).

Companies have an interest in authenticity because it is central to the sales process. An authentic product is associated with 'genuineness' and quality and the resulting appreciation (cf. Dietert 2018, p. 48). A close relationship with the company and the brand creates trust and, for the associated products, an authenticity of its own. This effect positions the economic actors far in the sphere of the social and interpersonal. Accordingly, the topic is not only discussed in marketing but has two well-known predecessors in intellectual history. One is Marx's thesis of the fetish character of commodities (cf. Marx

---

The translation of this chapter was done with the help of artificial intelligence (machine translation by the service DeepL.com). A subsequent human revision was done primarily in terms of content.

1956). Marx describes the capacity of products to take on social properties beyond their use-value. They acquire these properties through their processing and reflect them to the consumer. The commodity thus becomes a "social hieroglyph" (ibid., p. 88). It contains coded messages by which certain value relations immanent to the system are conveyed. Identifying such "metaphysical[s] quibble[s] and theological[s] muck" (ibid., p. 85) opens up the possibility of exploring a social phenomenon, such as authenticity in relation to products. Secondly, Guy Debord's (1996) approach stands, after capitalism turns any authentic human experience into a commodity and tries to sell it to the same people. Debord calls this mechanism, fueled by advertising and mass media, the "society of the spectacle," which encompasses every area of life and makes the world unreal (cf. Heath and Potter 2005 for a more recent reception). Here it becomes clear that inherent in the market system is the power to generate its world of perception by means of its products. To anticipate: this control of perception is the most important ingredient for companies to generate their authenticity for their products.

7.1 What Is Authenticity?

Authenticity goes back to the Greek term *authentikós* and means *genuine* or *vouched for*, having a relationship to the originator or original idea of a thing. It also denotes a contrast to deception. Overall, it describes a positively judged quality, a criterion of quality. Authenticity is desired and must be acquired. It creates legitimacy and trust once it is achieved. In the economic context, it thereby enables an ongoing and effective exchange relationship (cf. Zapf 2017).

Economically used authenticity is created by a triad, following Saupe (2012):

1. an *expectation of* the customer that arises from engagement with the product,
2. the direct *experience of* the product (external appearance) and
3. the *peculiarity of* the product, which is supported by a well-known story.

In religion, authenticity has the function of determining whether certain objects, actions or persons are part of the religion or not (cf. Radde-Antweiler 2012, p. 89). Religious institutions and their religious experts can define what is considered authentic by controlling the history of the issue in question through their authority and thus shaping the subjective expectation and direct experience of the followers. The same is true of authenticity in economics: the idiosyncrasy of an object is relative and cannot be measured by any *actual substance*. Instead, it can be influenced by institutions or an authority. This concerns the aspect of *ownership that is* intersubjectively shaped by history. Individual *expectations* and *experience of* the product are influenced by it.

Digitization makes it easier for companies to use their own story to influence the expectations and experiences of customers and thus generate their authenticity. This is due to the fact that an increasing number of products are purchased but are no longer tangible. For example, software as a download, digital media, such as music, books or films. You do not have to hold a CD in your hand to listen to music. You do not even have to own the music. Access authorization and the smartphone are enough to consume it. Thus, habituation in dealing with virtual products occurs. A market emerges that trades virtual objects with real money. Without the expectation of acquiring a tangible product - as the example of a computer game in which virtual equipment or weapons can be bought shows (keyword: *Loot Boxes*).

The pure experience is enough. This has consequences for product expectations in the analogue world. It is not so much the substance of an object that is decisive, but its story. An example: Tesla sells different battery capacities in its electric cars: 60 and 75 kWh, the latter for US$9000 extra. Only in both cases, the identical battery is installed. Only when buying the 'bigger battery' the higher capacity is unlocked via the software. This became known when Tesla remotely unlocked the larger range for all drivers near a hurricane to get out of harm's way. Subsequently, the old capacity was restored (cf. Kopf 2017). For our topic, a different aspect than this help in need is interesting. The naïve assumption would be: a battery cannot have several capacities at the same time, but is fixed in its peculiarity to one capacity. This is no longer true - the battery has the capacity ascribed to it by software and marketing. The customer's expectation and experience can no longer be created without the company's help: who understands the structure of a modern vehicle battery? So the company generates its authenticity by spinning a specific story and thereby shaping the customer's expectation and experience.

It is noteworthy that companies can also draw on existing perceptions of authenticity and reformulate them in their interest. This from the consideration of maintaining the existing willingness to pay for a product while the product itself becomes cheaper. Examples of this can be found in the food industry under the heading 'Umfruchten', as the Verbraucherzentrale Hamburg indignantly notes (2015). In cranberry fruit yoghurt, grape pieces are used instead of cranberries. Because grapes are cheaper and give a similar mouthfeel. The customer expects the consistency of a cranberry. The yogurt's eating experience meets that expectation; you cannot feel or taste the difference, thanks to the food engineers' ingenuity. The story told around the yogurt does the rest. The authenticity adopted means an increase in profit for the manufacturer by the difference in the price of grapes and cranberries. Other examples include whiskey that, thanks to clever blending and additions, tastes as if it has been aged for 20 years, even though it has only been in the barrel for a fraction of that time. Or 'analogue' cheese that contains little or no milk. If no difference is noticed when drinking and eating these products—are not they just as authentic as their 'real' counterparts?

7.2 Disney and Audi: Authenticity Brings Sales

A visit to Disney's *Epcot* amusement park in Florida shows just how well authenticity production works. More precisely: the *World Showcase* does, which makes it possible to experience the special features of different cultures. There are mainly restaurants (*Germany Pavilion*: Schnitzel), shops equipped according to the respective culture (cuckoo clocks made in China) and rides. The facade of the building is a mixture of half-timbering and knight's castle, slate roof and geraniums, most of it made of plastic. A look behind the facade reveals a huge room decorated in the style of Oktoberfest with beer benches. Everything you think you know about a culture brought to life in 2 min. Disney succeeds in creating an idiosyncrasy in the visitor through a story (the aforementioned interpretation of German culture) that frames the expectation and experience of a foreign culture in a way that creates the illusion of a visit to Germany. The twist with authenticity, however, takes place outside this hall, at a souvenir stand in front of plastic Germany. There, the visitor discovers a mug with a large Mickey head printed on it. Around it written in proud letters: AUTHENTIC DISNEY. In an environment that, to the uninitiated observer, radiates *the opposite of* authenticity in every single aspect, this mug ensures that beyond the special authenticity of the 'Germany visit', the entire Epcot product takes on the authenticity of *its own* ("Authentic Disney"). The visit becomes an inimitable, genuine Disney experience.

A second example illustrates the potential of digitalization for authenticity production: the Audi sound generator. Modern engines are supposed to be affordable and quiet. However, this eliminates an important feature of sporty combustion engines, the sound. Vehicle manufacturers are taking note. Using loudspeakers near the rear silencers, so-called sound actuators, certain frequencies of the engine sound are amplified and individual aspects are emphasized. According to Audi, this creates an "authentic (!) soundscape" (Audi 2019) that can be changed or turned off at the touch of a button. Individual models can also direct the sound only into the passenger compartment or prevent noise emissions altogether by counter-sound. However, it is usually the outward appearance that is advertised. The authenticity of the engine sound comes into question when bystanders cannot recognise the sound. In Audi-speak, "impress your surroundings" (ibid.) with the speakers.

The three-step of expectation-experience-ownership and the concise story told by the company are also applied here. The car buyer expects the corresponding sound from his powerful vehicle with an internal combustion engine, as he has been used to and internalized since childhood. The direct experience corresponds to this expectation. The vehicle sounds as expected. The authenticity of the sound is carried by the reference to the sporty engine sound of traditional internal combustion engines, which produce the sound through elaborate and artistic mechanics, the curvature of exhaust pipes, flaps and coils. Audi reformulates this authenticity with speakers and sound chips and sells the sound experience to the customer as an authentic soundscape, as described on their homepage.

7.3 Authenticity Without Secrecy

Disney sells its version of Germany, Audi its version of a V12. Facebook and Apple know what the most important personal photos and events of the year were. Authenticity is what is billed, advertised and offered for sale. Authenticity is divorced from the personal secret and the secret of things; on the contrary, it legitimizes itself only through publicity. It is not the peculiarity of a product that is decisive, but its design and the surrounding communication. The consumer's expectations and experience are based on this. This effect is intensified by virtual products that are beyond physical reach. The customer believes the authenticity because he cannot grasp it.

Impressive shopping experiences are created. The strange composition of one's photos is touching. The simulated world trip (Epcot) or the V12 sound (Audi) beguile the senses. Tesla pulls 100 km range out of its hat and saves lives. The customer is picked up on all levels of perception and can immerse himself in the product. The immersive consumer experience: a remarkable achievement of the digital economy.

References

Audi (2019): *Motorsoundsystem*. Homepage Audi Schweiz. https://www.audi.ch/ch/web/de/kundenbereich/audi-original-zubehoer/layer/motorsoundsystem.html (11.12.2019).

Debord, Guy (1996 [1967]): *Die Gesellschaft des Spektakels*. Berlin: Edition Tiamat.

Dietert, Anna-Christina (2018): *Erfolgssicherung von Marken durch Authentizität*. Heiderlberg: Springer.

Heath, Joseph; Potter, Andrew (2005): *The Rebel Sell. Why the Culture can't be jammed*. Chichester: Capstone.

Kopf, Dan (2017): *Tesla intentionally makes some of its cars worse, and it's good for everybody*. https://qz.com/1074721/tesla-intentionally-makes-some-of-their-cars-worse-and-its-good-for-everybody/ (10.12.2019).

Marx, Karl (1956 [1859]): Der Fetischcharakter der Ware und sein Geheimnis. In: Marx, Karl; Engels, Friedrich (Hg.): *Werke. Band 23. Herausgegeben vom Institut für Marxismus-Leninismus beim ZK der SED*. Berlin: Dietz-Verlag, S. 85–99.

Radde-Antweiler, Kerstin (2012): Authenticity. In: Campbell, Heidi A (Hg.): *Digital religion: Understanding religious practice in new media worlds*. New York: Routledge, 88–103.

Saupe, Achim (2012): *Authentizität*. Clio-online. https://zeitgeschichte-digital.de/doks/files/263/docupedia_saupe_authentizitaet_v2_de_2012.pdf (17.10.2019).

Zapf, Chr. Lucas (2017): „Strategisches ethisches Risikomanagement (SERM). Bewältigung von Legitimitätsrisiken in der Praxis", *ZRFC Zeitschrift für Risk, Fraud & Compliance*, 5 (17), 203–8.

Zapf, Lucas (2020): „Authentizität - der Glaube macht's", *Neue Wege*, 1/2 20 16–20.

From Ethical Considerations to Proposed Legal Solutions: An Afterword by *Bertil Cottier*

After reading this book, pessimism sets in. From now on, the citizen is completely helpless: none of his gestures, his attitudes, his preferences, his intentions, his opinions, his convictions, as well as his doubts and concerns can escape the large-scale surveillance made possible by the new technologies of information processing, starting with *big data and artificial intelligence.* Nowadays, powerful profiling and tracking tools are available. The state, as well as companies, are not afraid to make full use of them, for better or for worse. As much as people call for restraint, point out the egregious abuses and condemn the violation of personal freedom: confidentiality is over. All-round surveillance is a reality. As the two authors of this study have pointed out in conclusion: "Digital omniscience is (…) on its way to being established. The secret private is (…) gone".

As a lawyer, I too draw a similarly negative conclusion, even if at first glance there is no lack of instruments to protect privacy. At the international level, both the International Covenant on Civil and Political Rights and the European Convention on Human Rights devote a specific provision to the protection of privacy. The same is true at the national level for most modern constitutions, such as those of Switzerland or South Africa. These general basic provisions have been fleshed out by legislation on data protection, the primary aim of which is to enable citizens to control the use of their personal data. The first such legislations—the law of the German state of Hesse in 1970 and the Swedish law of 1973—were adapted to counter the threats to privacy posed by the first computers and, in particular, by their ability to create and network large personal files. To date, more than 140 countries have adopted data protection laws (the latest to date being Kenya, Egypt and Uzbekistan).

Although the scope of these different laws varies—today, European laws are undoubtedly still the strictest—they all contain some guiding principles that regulate in detail any handling of personal data, such as its collection, storage, processing, transfer and destruction. In short, the data subject must know how his or her data is used (the so-called principle of *transparency*). In addition, the processing of data must be limited to the information that is indispensable to achieve the desired objective (principle of *proportionality*). Furthermore, data should not be used for a purpose other than that for which it was collected (*purpose* limitation principle). The data subject has the possibility to oppose processing that is contrary to him or her (*right to blocking*) and to know who is processing data about him or her and for what purpose (*right of access*). There is also the specific protection of sensitive data, in particular political opinions and religious beliefs, as well as health data and information on sexual orientation. Finally, most data protection legislations establish a specialised body to promote and implement the protection standards. Although the composition of this body varies from country to country—either a commissioner or a commission, or a combination of both—it has the status of an independent authority everywhere. Conclusion: legal means exist to prevent violations of privacy. One must, therefore, ask why they prove ineffective.

There are several reasons for this ineffectiveness. First of all, there is the fact that the United States stands out for its lack of any general legislation on data protection. *No omnibus data protection law*! is a slogan often used by the American administration. Neither Democrats nor Republicans intend to subject businesses to strict data protection obligations. If there is a real need to intervene, it will be selective to address a specific serious violation (such as an employer monitoring social media activity or a security breach that caused the loss of personal data). Because of this lack of general data protection legislation, internet giants, from Google to Facebook almost all of American origin, have a free hand, so to speak, to create profiles of internet users on a grand scale. Regardless of where they are located.

The inability of the legal system to protect privacy is also the consequence of a "design flaw". If a person is to be granted a genuine right to "informational self-determination", then data protection laws must allow each individual to decide independently on the use of their personal information. Refusal to allow this or that service provider to collect personal information and then share it with commercial partners must be possible, as must consent. The weak point is that consent is rarely a conscious decision (as the legislator wrongly assumed), but mostly a "self-inflicted immaturity", to take up the very apt expression of the authors of this publication. Too often, users give this consent too quickly and recklessly because they are in a hurry to enjoy the benefits of the services and are not eager to read through endless terms of use forced upon them. Equally negligent are the countless Internet users who unrestrainedly flaunt themselves on social networks and thus help commercial profiling agencies to make a big profit.

The third inefficiency factor is the low number of complaints brought before the courts or the data protection authority. Because privacy violations rarely involve financial damages, affected individuals are often reluctant to engage in protracted legal battles to assert their rights. Not everyone is as committed as Max Schrems, an Austrian privacy activist who brought down the EU-US personal data transfer agreement (the so-called *Safe Harbor* agreement) on the grounds that his data, once on the other side of the Atlantic, could be secretly transferred to American intelligence agencies. Nor is everyone like Costeja González, a small business owner from Barcelona who went all the way to the Court of Justice of the European Union in the name of a *right to be forgotten to* force the search engine run by Google to stop referring to financial difficulties he had a few decades ago. Gonzalez and Schrems are exceptions. Mostly, affected individuals just fold their arms, discouraged by the prospect of protracted, exhausting, and risky litigation (one must also remember that the litigation loser must pay for the costs).

The last cause of the ineffectiveness of data protection is the most alarming. It has become topical again, especially with the advent of *big data and artificial intelligence*. With these new techniques of information processing, based on powerful statistical calculations, we have lost the ability to make the legally crucial distinction between personal data and non-personal data. *Big data* processing algorithms make it possible to derive personal data by correlating enormous amounts of non-personal data. The processing of non-personal data escapes data protection law—this can only be applied to the processing of personal data.

Due to the four reasons explained above, traditional data protection laws struggle to counter large-scale surveillance and data tracking. Could a revision of the relevant texts address these shortcomings? Yes and no.

It is true that the modernization of data protection laws is underway. The trigger for this modernisation was the entry into force of the European Union's General Data Protection Regulation (GDPR), which created uniform privacy standards for all 27 Member States. This text contains numerous innovations, such as the strengthening of the quality of consent (each person must be able to express his or her opinion on the data processed and the purpose of the processing by means of clear and comprehensible information), the right to the erasure of useless or obsolete data, or even the obligation to take data protection into account from the moment a service is designed (*privacy by design*). Two novelties deserve to be mentioned. First, the restrictions of the so-called "profiling". This new concept includes

> any automated processing of personal data which consists in using such personal data to evaluate certain personal aspects relating to a natural person, in particular, to analyse or predict aspects of that natural person's performance at work, economic situation, health, personal preferences, interests, reliability, behaviour, location or change of location (Art. 4(4) GDPR).

In practice, it is mainly abuses of data mining that are in the sights of European legislators. The second important novelty is the extension of the scope of application of the legal provisions of data protection. No longer will only companies located in the territory of the European Union have to comply with the rules, but also those that engage in activities that are likely to affect European citizens (regardless of the company's location). This extension should eventually make it possible to take action against the American Internet giants, which until now have always cited their non-European location as an argument to escape data protection requirements. It seems to be working: in January 2019, Google was fined 50 million euros by French data protection authorities for failing to provide its users with reliable information about their right to object.

As ambitious as the revision is, it seems insufficient to contain the emerging phenomenon of *big data* and its almost unlimited possibilities for surveillance and data tracking. Thus, at first glance, the concept of profiling seems revolutionary. On closer inspection, however, it is less so, because it still only applies to personal data. All the more so, the European General Data Protection Regulation has a serious flaw. It is based on the fundamental right of "informational self-determination", a right that was enacted by the German Federal Constitutional Court in 1983, at a time when the threat to privacy was very different from today.

With *big data,* the right to "informational self-determination" has become an illusion. If one really wants to protect privacy in the context of ubiquitous profiling, a new basis must be created. The approach of "empowerment" of the data subject, which has been predominant until now, should be replaced by an approach based on the risk of big data processing. In other words, the attention of the legislator must focus less on what the data subject wants and much more on the risks posed by the data processing. The legislator must address the public and private operators by telling them exactly how to prevent abuses of privacy.

But that is not all: this change of direction must be accompanied by a veritable education in the conscious use of "*privacy*". Operators must be made aware of the right to privacy. Indeed, the rules that protect privacy are much more easily respected when they have the support of the people concerned. So let the authors of this work be warned: we are counting on them to spread this message.

Lausanne/Lugano in February 2020

GPSR Compliance

The European Union's (EU) General Product Safety Regulation (GPSR) is a set of rules that requires consumer products to be safe and our obligations to ensure this.

If you have any concerns about our products, you can contact us on

ProductSafety@springernature.com

In case Publisher is established outside the EU, the EU authorized representative is:

Springer Nature Customer Service Center GmbH
Europaplatz 3
69115 Heidelberg, Germany

www.ingramcontent.com/pod-product-compliance
Lightning Source LLC
LaVergne TN
LVHW021334080526
838202LV00003B/169